HOMEMADE

80+ HOUSEHOLD ESSENTIALS TO INSPIRE YOUR EVERYDAY

HOMEMADE

80+ HOUSEHOLD ESSENTIALS TO INSPIRE YOUR EVERYDAY

Eleanor Ozich

PENGUIN BOOKS

Contents

Introduction

Firstly, I'd like to say a warm, heartfelt hello! I'm so pleased that you've picked up this book, and I absolutely cannot wait to share my love of cooking and eating with you.

I first started sharing recipes on my blog about six years ago, and since then, I've learnt that the key to creating healthy, exciting food is to have plenty of basic recipes which bring it all together. As a mother of three, I certainly don't have all the time in the world to spend in the kitchen – I'm always trying out simpler ways to cook, keeping methods as fuss-free as possible, yet still focusing on an insanely delicious outcome.

I love to eat in a way that makes me feel light, energetic and nourished. This doesn't mean I spend hours on end making fancy or extravagant dishes – quite the opposite, in fact! My simple approach to cooking means that substantial and delicious meals can be whipped up in a flash, so long as I have my essential go-to recipes up my sleeve.

Within the pages of *Homemade* you'll find an abundance of recipes that inspire you to try something new. It's a book that covers all kinds of back-to-basic methods using simple, everyday ingredients that don't cost the earth.

Perhaps you might like to bake a wholesome loaf of almond bread, make your own honey-scented oatcakes, learn the basics of culturing yoghurt, or be intrigued by rosemary-infused caramel sauce? More often than not, we buy these pre-made foods from the supermarket, without knowing how satisfyingly easy they are to make at home. However, once you have a go at making such basics from scratch, you'll find they're so much more delicious than the plastic-packaged counterparts.

Honestly, you'll begin to know these recipes off-by-heart they're so easy to incorporate into your daily rhythms. You might even begin to notice your fridge and pantry shelves filling up, and then wonder how you ever cooked without these homemade basics.

Homemade is a fun, satisfying and delicious way to eat, and one that considers your health, and the environment, too. My hope is that you will use these recipes to inspire your very own colourful meals, while also bringing a new sense of passion and creativity to your kitchen.

Love, Eleanor x

01: Kitchen essentials

In this first chapter, I share a breakdown of the simple, everyday provisions that I've come to love and depend on. These recipes for wholesome supermarket substitutes are not only versatile when it comes to creating quick, healthy meals, but they're much more cost-effective, and can make your meals go so much further.

Over the years, I've also come to learn that a well-equipped, organised kitchen is essential for feeling inspired to cook. There's no need for fancy equipment or the latest gadgets, rather, I prefer to keep things simple and invest in quality tools that are both practical and beautiful on a daily basis.

You'll also see a basic guide to sprouting, as I find this simple art to be an important piece of the puzzle when it comes to healthy eating. Moreover, sprouting can be a fun and nutritious way to bring new life to dry ingredients in your pantry.

Pantry staples

Nuts and seeds are a brilliant way to provide your body with more vitamins and nutrients, and they are also high in protein and packed with good fats. Add a variety to your salads, oatmeal, muesli and sweet treats, or use chia seeds in place of eggs in baking. My favourite nut- and seed-filled recipe in this book is my super seeded crackers on page 91. They're effortless to make and incredibly delicious. You simply have to give them a go!

Coconut cream and coconut milk are wonderful options in place of dairy in baking, smoothies and pancakes; try my triple coconut waffles (page 113). Fantastic, also, for making your own coconut yoghurt (page 59). Try to choose brands without any added thickeners or additives.

Grains such as quinoa, buckwheat, pearl barley or wild rice are a simple way to include more satisfying texture in your cooking. I find a humble jar of grains is like an old friend I can fall back on – they help me to create an impromptu-style dish out of a seemingly empty fridge. Additionally, grains are cost-effective and filling, particularly if you buy in bulk.

Free-range, organic eggs make a nutritious breakfast, lunch or dinner. The humble egg is considered to be packed full of all kinds of goodness, which makes me happy as our family goes through at least two dozen a week. Always choose eggs that are free-range or organic, if possible, not only for ethical reasons, but because they'll be better for you.

I'll always have a selection of **unique flours** on hand that are gentler on the digestive system, and are much more interesting to eat. These include spelt flour, oat flour, buckwheat flour and quinoa flour, as well as plenty of raw nuts to grind into nut flour as I need it. I also keep a large jar of my own blended gluten-free mix, which you'll find on page 35. When you welcome a little variety into the flours you use, you'll be pleasantly rewarded with all kinds of wholesome baked goods.

Legumes such as lentils, butter beans and chickpeas are the perfect base for a simple, nutritious meal. They're packed full of protein, making them fantastic for vegetarians, as well as being high in fibre and long-lasting carbohydrates. If using dried, always ensure you soak them for at least 12 hours before cooking so they're gentler on your digestion. I'd also like to take this opportunity to mention my roasted garlic and butter bean hummus on page 133 – it's a goodie!

Natural sweeteners such as maple syrup, honey, coconut sugar and rapadura sugar are brilliant for more wholesome baking and sweet treats. I personally believe the taste is far superior compared to regular white sugar, plus natural sweeteners prevent you from spinning into a sugar high as they are known to give you a more sustained and slower energy release.

Rolled oats, flaked quinoa and brown rice flakes are humble, low-cost ingredients that we should enjoy in our morning oatmeal, muffins and muesli bars. I often use rolled oats in baking instead of flour or ground nuts as it is a more cost-effective option. Simply grind the oats in your food processor until you have a flour-like consistency.

Extra virgin coconut oil is excellent for cooking as it has a high smoke point, and it is great in raw desserts or treats as it solidifies when chilled, creating an astoundingly rich and delicious texture. I'm all about incorporating plenty of good, healthy fats into our diet, and I love coconut oil as it supports brain function, immunity and healthy skin.

Other various flavour bombs include olives, sundried tomatoes, capers, Dijon mustard, organic tomato paste, and bottled organic chopped tomatoes for quick, delicious sauces. I also have a shelf dedicated to dried herbs and spices, as well as various vinegars and extra virgin olive oils for drizzling. These are the ingredients that set my taste buds dancing, and get me excited to create vibrant food.

Handy tools and equipment

Each Sunday, I'll spend the afternoon in the kitchen preparing a good selection of basic recipes so that I'm all stocked up and prepared for the week ahead. The best way to make this as simple as possible is to have a handy collection of tools and equipment that makes the job a whole lot easier.

We all have our own idea of what is necessary in the kitchen – it is, after all, a uniquely personal space! However, I thought you might be interested in the tools that I love and use on a daily basis.

Our **slow cooker** is running almost year round and holds fragrant, steaming bone broth whenever we need it. My family and I sip on it throughout the day as it's incredibly nourishing and healing for gut health, as well as being fantastic for adding more depth to my cooking. Soups, stews, casseroles and roasts are all lovely ways to make use of your own homemade bone broth, and you can find the basic method on page 157.

My favourite kitchen tool is undoubtedly my **cast-iron skillet.** It may seem like a bit of an old-fashioned choice, but I believe this dependable cookware is a must in any modern kitchen. Given plenty of love and care, cast iron can become a quiet treasure on your stovetop that will last a lifetime and help you to create many delicious meals. The same goes for my **Dutch oven pot**, which can be used both on the stovetop and in the oven for slow-cooked masterpieces and sourdough bread-making.

Throughout this book, you'll find that I use both my **blender and food processor** in the majority of my recipes, and for good reason, as it makes life ridiculously easy! Particularly handy for baking recipes, whipping up truffles and raw treats, and blending flavour-packed sauces and dips.

Whether you're a seasoned cook or just starting out in the kitchen, a selection of **good knives** is essential to making your cooking journey easier and more enjoyable. I have two Pallarès carbon-steel chef-style knives which have become the most important and beloved tools in my kitchen – they're a little harder to care for as they can rust easily, but I just love the way they feel in my hand, and the rustic patina that's formed over the years.

I also use a few smaller knives from French brand Opinel. Their nimble size and narrow blades help to slice smaller ingredients with more accuracy, such as cutting tomatoes, peeling fruit or hulling strawberries.

I have quite a collection of **eco-friendly storage options**, including glass jars and stainless steel containers. Not only are they prettier, and a healthier way to store food, but they last much longer, too. The same applies for wooden utensils, which I've never had to throw out – I can't say the same about plastic versions.

Cotton and linen storage bags are also a fabulous earth-friendly companion to take along with you to the local farmers' market or grocery store. The best part is that you can use them over and over again, and pop them in the washing machine when necessary. Linen dish covers can be used in place of cling-film, while beeswax food wrap is another idea to keep your food fresh without plastic – see my method to make your own on page 218.

A calm, inspiring kitchen

Keeping my kitchen as clean and clutter-free as possible helps me to feel relaxed and inspired when it comes to cooking. When I'm feeling calm in the kitchen, I tend to put far more love into my food, and the outcome is infinitely more delicious.

One of the easiest ways to keep your kitchen looking spick and span is to keep the counter space as clear as possible. Once you have tucked everything away neatly, select a handful of everyday items that you're especially fond of and arrange them on your bench or kitchen shelf in a simple, yet beautiful display. This could be a ceramic vase filled with beautiful utensils, a cast-iron pot sitting on the stovetop, and perhaps a small arrangement of rustic, yet elegantly worn, cutting boards.

It's these little, thoughtful touches that will add a sense of character to your kitchen, and make it feel like a sacred, beautiful place, which you'll naturally want to spend more time in. Lovely music playing in the background also helps to create a beautiful cooking environment.

Some words on sprouting

Sprouting might seem like a new health trend to some, but, in fact, this traditional technique has been practised for centuries. The simple preparation begins by soaking the ingredient – whether it be grains, legumes or seeds – in water to break down the outer shell. The ingredient is then drained, rinsed, and kept moist in a jar to encourage growth. This basic process helps to break down the starch, allowing the ingredient to release more nutrients, making it more easily digestible.

Simple as can be, the method of sprouting doesn't require any fancy equipment and gives you quick gratification as the process typically only takes a few days. You'll also love how these soft and chewy sprouted morsels can add intriguing texture and taste to salads, baked goods, warming stews and more.

To keep things easy, I've called the ingredient in these instructions 'grains', however, you can apply the same method to sprout most types of legumes and seeds that are raw, including quinoa, spelt and buckwheat.

Place the grains in a sieve or colander, rinse well and drain.

Next, place the grains in a glass jar, bearing in mind that they will increase in volume (usually 3–4 times the original amount). Cover with filtered water by 5–6cm, place a sprouting lid or double layer of muslin on top, then secure with a rubber band. Allow to soak for 24 hours.

Transfer the grains into a sieve or colander, rinse and drain well.

Place the grains back into the jar, cover with the muslin, and place somewhere out of direct sunlight.

Once every day, pour water into the jar and swirl it to evenly rinse all the grains. Drain the grains in a sieve and rinse, and then transfer them back into the jar.

Repeat this process for 2–5 days. You will know when they are ready as they'll have little tails popping out.

Once they have sprouted to your desired length, rinse and drain the sprouted grains, then place them in a container lined with a paper towel, and store them in the fridge for up to a week or so. If they start to smell funny or look slimy, discard them.

02:   Pantry basics

The ingredients in my kitchen often change depending on the season, however, I'll always have some of my favourite store-cupboard basics to line the pantry shelves so that I'm never caught off-guard.

Within this chapter, you'll find a colourful spectrum of essentials which I use to create simple, joyful meals and particularly nice things to drink. I'm always looking for more flavoursome ingredients I can grab to add more depth and excitement to my dishes, and it's these staples that help me to keep things fresh and healthy in the kitchen.

From delicately spiced hot cacao mix to fragrant dried stock, you'll find brightly coloured cordials sitting amongst rows of preserved fruit, glowing beautifully in jars. These lovingly made pantry basics will bring joy to those days when you're feeling least inspired. Simply throw open your cupboards, and allow the ingredients to reveal their magic.

Golden ghee

I love the weekly ritual of preparing a jar of golden and delicious ghee. Also known as purified or clarified butter, ghee has been used in Ayurvedic medicine for thousands of years and is known to nourish and regenerate body tissue, cleanse toxins, and aid in brain function, while also boosting the immune system and promoting general vitality. Ghee is a brilliant alternative to most other vegetable oils, and adds an indulgent toasted flavour to your food.

500g unsalted butter, preferably organic
large piece of muslin or cheesecloth

Slice the butter into cubes and place in a large pan over medium heat. Once the butter has melted, turn the heat to low and allow the butter to gently bubble away for 15–20 minutes. Your kitchen will be filled with the most divine, toasty aroma, similar to popcorn or the smell of pancakes cooking on a Saturday morning.

You'll know the ghee is ready once the butter has turned a lovely dark golden colour, and a foam has appeared on the top. Remove from the heat and carefully pour through a muslin- or cheesecloth-lined sieve placed on top of a glass jar.

Store in a cool, dark place for up to 3 months.

Fragrant dried stock

Most store-bought stock powders are filled with additives, preservatives and flavourings that I'd rather do without. So I decided to come up with my own homemade version using simple ingredients from my pantry. The result is a lovely stock that's fragrant and delicious, perfect for adding to soups, stews and casseroles for a rich, deep flavour.

Makes 1 cup

¼ cup Himalayan or Celtic sea salt
⅓ cup brown rice flour
2 tablespoons dried onion powder
2 tablespoons dried garlic granules
1 tablespoon dried thyme
1 tablespoon dried rosemary
2 teaspoons dried porcini powder
1 teaspoon dried oregano
1 teaspoon mustard seeds or powder
1 teaspoon ground turmeric
1 teaspoon fennel seeds or powder
1 teaspoon ground black pepper

Place all ingredients in a food processor and blend until you have a fine powder. The dried stock will be a sunshine yellow colour thanks to the turmeric. Transfer to a small glass jar.

To use in cooking, dissolve about 1 heaped tablespoon of stock powder in 4 cups of water. I also like to sprinkle it over root vegetables before roasting.

Store in the pantry for up to 1 year.

Aromatic fennel salt

Tantalising and fragrant, this easy to make salt rub can be sprinkled on all manner of things, such as oven-roasted root vegetables, slow-cooked meat or grilled fish. Use sparingly, and add more as you desire.

Makes 1½ cups

⅔ cup coarse sea salt
½ cup fennel seeds
2 tablespoons pink peppercorns
2 tablespoons dried garlic granules
2 tablespoons dried rosemary

Place all the ingredients in a food processor and process until well ground. You could also do this in a pestle and mortar. Transfer to a medium-sized jar.

To use, lightly coat your meat, seafood or vegetables with olive oil, and sprinkle some of the seasoned salt over the top. Allow to marinate for half an hour or so before grilling or roasting.

Store the salt in a cool, dry place for up to 6 months.

Vanilla-scented coconut sugar with cinnamon

This is a great recipe to breathe new life into already scraped vanilla beans. There's an abundance of uses for this softly scented sugar: sprinkle the stuff over pancakes, waffles, grilled fruit, or even your morning bowl of oatmeal. I also suggest stirring a teaspoon into your cup of tea if you're feeling a little fancy.

1 cup coconut sugar
4 teaspoons ground cinnamon
2–3 vanilla beans

Combine all the ingredients and toss until well combined. Transfer to a medium-sized jar.

Store the jar in your pantry for up to 1 year.

Roasted hazelnut and mint dukkah

Dukkah is an Egyptian spice blend that provides a burst of piercingly intense flavour, and an intriguing crunch when added to dishes. It's quite wonderful sprinkled over grilled eggs, roasted cauliflower, or spooned into silky potato mash to smarten it up. Dukkah is also brilliant with a loaf of sourdough bread and a lovely olive oil for dipping.

Makes 1½ cups

¾ cup sesame seeds
½ cup hazelnuts
¼ cup fennel seeds
4 tablespoons cumin seeds
4 tablespoons coriander seeds
2 tablespoons dried mint
1 tablespoon smoked paprika
1 tablespoon chilli flakes
1 tablespoon pink peppercorns
1 tablespoon sea salt

Combine all the ingredients and dry toast them in a large skillet over low heat, stirring with a wooden spoon until fragrant.

Allow to cool, then transfer to a food processor and pulse until you have a coarse crumb consistency. Transfer to a medium-sized glass jar.

Store in the pantry for up to 6 months.

Oat, coconut and chia seed cookie mix

These little honey-coloured oat biscuits are mercifully less sweet than the majority of biscuits. A little spice or zest could work nicely with the delicate coconut notes, and they are perfect for tucking into the children's lunchboxes. This pre-made mix also makes a wonderful gift.

Makes about 18 biscuits

For the cookie mix
1½ cups rolled oats
1¼ cups spelt or
    buckwheat flour
⅔ cup rapadura or
    coconut sugar
½ cup desiccated coconut
3 tablespoons chia seeds
½ teaspoon baking soda
pinch of sea salt

To bake
¼ cup melted coconut oil
    or unsalted butter
2 tablespoons water

Layer the cookie mix ingredients in a jar, pop the lid on, and store in the pantry until you are ready to bake.

Preheat the oven to 170°C and line a tray with baking paper.

Pour the jar of ingredients into a mixing bowl, add the melted coconut oil or butter and water, then stir to combine. Allow the mixture to rest for 5 minutes.

Using your hands, roll the mixture into small balls about the size of golf balls and place 2cm apart on the baking tray. Press down lightly on each ball to flatten.

Bake for 12–15 minutes, or until golden, then allow to cool on the baking tray. The biscuits will keep in an airtight container or jar for up to 1 week.

All-purpose gluten-free flour mix

Unlike the majority of gluten-free flours out there that often have a peculiar texture in baking, this easy, five-ingredient all-purpose flour mix produces consistently delicious results. You'll also be happy to know that in most types of recipes, this blend can be used in place of regular flour.

2 cups brown rice flour
2 cups sweet rice flour
2 cups white rice flour
2 cups tapioca flour
1 tablespoon xanthan gum*

The xanthan gum can be left out, although it works as the gluten replacer in recipes where yeast is needed, such as pizza dough or bread

Place all the ingredients in a very large mixing bowl, and whisk together for 1–2 minutes until very well incorporated.

Store in a large airtight container or glass jar in the pantry for up to 1 year.

Preserved lemons with black pepper

Preserved lemon is exquisite with roast chicken, added to luscious mayonnaise (page 131), stirred into slow-cooked lamb, or sautéed with vegetables for a pop of flavour. Also try preserved lemon with softened butter and fresh herbs, and smear all over grilled fish.

½ cup sea salt, plus more if needed
8–10 organic lemons
1 tablespoon peppercorns
extra freshly squeezed lemon juice, if needed

Place 2 tablespoons of the salt in the bottom of a large sterilised jar (see page 55 for sterilising instructions).

One by one, prepare the lemons in the following way. Cut off any protruding stems, then slice the tip off each lemon. Slice the lemons as if you were going to cut them in half, lengthways, starting from the tip, but do not cut all the way down – keep the lemons attached at the base.

Make another slice in a similar manner, so now the lemon is quartered, but still attached at the base. Pry the lemons open, and generously sprinkle salt all over the insides and outsides.

Stack the lemons in the jar, alternating with the peppercorns, squashing them down as you go to extract the juice. Ensure the lemons are fully immersed in the lemon juice, adding extra if needed. Top with a generous sprinkle of sea salt.

Seal the jar and pop on a shelf for 3–4 weeks, turning the jar upside down occasionally. Once opened, you can store the lemons in the fridge for up to 1 year.

Blackberry, lime and ginger cordial

To me, homemade cordials are a fabulous opportunity to create something vibrant, zippy and refreshing to drink. More often than not, we settle for the sugar-filled options from the supermarket, but why would you when you can make your very own prettier, fruit-filled versions. To serve, top with sparkling water, slices of lime and a sprig of thyme.

Makes about
2 cups of cordial

2 cups coconut sugar
2 cups water
2 cups blackberries, fresh or frozen
1½ cups lime juice
3 thumbs of ginger, finely sliced

To serve
sparkling or still water
ice cubes
slices of lime
sprigs of thyme

Combine all ingredients in a saucepan over medium heat. Bring to a soft boil, and then simmer over low heat until reduced by half, about 20–25 minutes. Allow to cool, pour through a fine sieve, and transfer to a sterilised glass bottle.

To serve, pour 30–40ml in a glass, depending on the size, and then top with sparkling or still water, 2–3 ice cubes, slices of lime and a sprig of thyme.

Turmeric lemonade cordial

Tart and sweet, this brightly hued cordial is the essence of summertime. This recipe yields about 2 cups of syrupy goodness, although I'll often double the recipe and pour it into lovely glass bottles to give away as gifts.

Makes about
2 cups of cordial

2 cups lemon juice
2 cups coconut sugar
2 cups water
3 thumbs of turmeric, finely sliced

To serve
sparkling water
ice cubes

Combine all ingredients in a saucepan over medium heat. Bring to a soft boil, and then simmer over low heat until reduced by half and syrupy, about 20–25 minutes. Allow to cool, pour through a fine sieve, and transfer to a sterilised glass bottle.

To serve, pour 30–40ml in a glass, depending on the size, and then top with sparkling water and 2–3 ice cubes, if desired.

Vanilla matcha mix

I switched out my morning coffee for matcha about a year ago, and now I'm utterly and hopelessly hooked. Matcha is super trendy these days, and for good reason, as it makes you feel fantastic (minus the jitters and crash of coffee) and is good for you, too.

This high-quality, powdered green tea contains a small amount of caffeine, which creates a calm, long-lasting alertness. It's also full of antioxidants, and holds many other health benefits. Serve with homemade nut milk.

Makes 1½ cups

½ cup culinary or ceremonial grade matcha powder*
1 cup lucuma powder*
1 tablespoon pure vanilla bean powder

To serve
1 cup nut milk (page 75)
1 tablespoon coconut oil
honey or maple syrup to serve, optional

Combine the matcha, lucuma and vanilla bean powders and transfer to a medium-sized jar.

To prepare, gently heat the nut milk until steaming, and then pour into a blender, along with 2 teaspoons of matcha mix, the coconut oil and a dash of honey or pure maple syrup, if using. Blend until creamy, then pour into a mug and enjoy.

* Available at organic foodstores

Delicately spiced hot cocoa mix

There is something quite magical about a frosty winter evening, the kind where cold hands are warmed with mugs of steaming hot cocoa, and blankets are draped accordingly. In this recipe, good quality cocoa or cacao powder is blended with golden coconut sugar, powdered coconut milk and a hint of spice to create a calming and enlivening hot cocoa mix. Simply spoon into a mug, and top with hot water – it couldn't be any easier.

Makes 3 cups

1½ cups good-quality cocoa or cacao
1 cup coconut milk powder or whole milk powder
¾ cup coconut sugar
1 teaspoon pure vanilla bean powder
1 teaspoon ground cinnamon
½ teaspoon ground nutmeg
¼ teaspoon ground cardamom
¼ teaspoon fine pink Himalayan sea salt

Sift the cocoa or cacao and milk powder into a medium-sized bowl. Add the remaining ingredients and whisk to combine evenly. Transfer to a medium-sized jar or airtight container.

To serve, spoon 2 heaped tablespoons of the cocoa mix into a mug, then top up with boiling water.

Store the dry mix in the pantry for up to 1 year.

Turmeric chai

Turmeric and ginger are known to have powerful anti-inflammatory properties, and this is possibly one of the most delicious recipes to make use of their medicinal magic. Subtly spiced with cinnamon and cardamom, I've also included coconut sugar and a dash of pure vanilla bean powder in the mix for a touch of sweetness.

Makes 1 cup

½ cup ground turmeric
⅓ cup coconut sugar
2 tablespoons pure vanilla bean powder
2 tablespoons ground cinnamon
1 tablespoon ground ginger
½ tablespoon ground cardamom

To serve
1 cup nut milk (page 75)
1 tablespoon coconut oil

Combine all of the dry ingredients and transfer to a small 250ml jar.

To prepare, gently heat the nut milk until steaming, and then pour into a blender, along with the coconut oil and 2 teaspoons of the chai mix.

Blend until creamy, then pour into a mug and enjoy.

Store the dry chai mix in the pantry for up to 1 year.

Salted maple buckwheat muesli

Each time I make this golden toasted muesli it never quite looks the same. That being said, I do love this muesli's changeable nature, and how I can endlessly switch out ingredients and it'll still magically turn out delicious. I've used buckwheat groats as the base on this occasion, as I love their delicate, earthy flavour, although wholegrain oats or quinoa flakes are also good options.

Makes about 10 serves

2 cups buckwheat groats
1 cup coconut flakes
3 tablespoons melted coconut oil
2 tablespoons almond butter (page 53)
2 tablespoons maple syrup or honey
2 teaspoons pure vanilla bean extract
½ teaspoon sea salt
1 cup dried fruit of your choice, I like figs, currants or blueberries
½ cup pumpkin seeds, or seeds of your choice

Preheat the oven to 180°C. Line a large tray with baking paper.

Combine all the ingredients except for the dried fruit and pumpkin seeds in a large mixing bowl. Mix until well combined, then spread the mixture out evenly on the prepared tray.

Roast for 15–20 minutes, turning the mix with a wooden spoon once or twice to ensure it browns evenly.

Leave to cool, mix through the dried fruit and pumpkin seeds, and transfer to a large glass jar or airtight container.

Store in the pantry for up to 6 months.

Sweet and salty toasted nut butter

Boasting the perfect balance of sweetness and saltiness, this nutty spread is blended to silky perfection. I slather it over sourdough, blend it into banana shakes, or simply eat it straight off the spoon. You'll also find it adds a luxuriousness to baking and is a wonderful filling inside fresh, gooey dates or dried figs. In fact, the more I daydream about toasted nut butter, the more I dream up ways in which this wholesome condiment can be enjoyed. And because it's ridiculously easy to make, well, why wouldn't you?

Makes 1 × 400ml jar

2 cups raw nuts of your choice
2 tablespoons coconut oil
2 tablespoons maple syrup
2 teaspoons pure vanilla extract
1 teaspoon ground cinnamon
½ teaspoon sea salt

Preheat the oven to 180°C. Line a large baking tray with baking paper.

Lay the nuts on the tray single file, then roast until fragrant, about 6 minutes. Remove from the oven, and allow to cool.

Transfer the nuts to a food processor along with the remaining ingredients, and process for 2–3 minutes until the nut butter reaches a smooth consistency. You might need to scrape down the sides a few times, or add another tablespoon or two of coconut oil.

Taste, and add a little more spice or sweetener if you desire. Transfer to a medium-sized glass jar. Store in the pantry for up to 6 months.

Preserved fruit with only a touch of sugar

This is an easy way to preserve most types of fruit, with only a touch of sugar. It works well for all varieties of stone fruit, berries, figs, apples and pears. The fruit will soften ever so slightly, while keeping its shape and structure. A fantastic method to have in your back pocket for when the fruit tree is plentiful.

fruit of your choice
1-litre preserving jars
coconut or rapadura sugar, 4 tablespoons for each 1-litre jar
lemon juice, if preserving apples or pears

To sterilise the jars, wash the jars and lids thoroughly, then place them in a large pot of boiling water, and boil for 20 minutes. Allow to cool for a few minutes, then, using tongs, carefully remove the jars and lids, and place them on a draining rack. Do not use a tea towel as this will contaminate the jars.

In the meantime, clean the fruit and, if needed, cut into thin slices, removing any pits or stalks. If you are using apples or pears, squeeze the lemon juice over the fruit and toss to ensure they are well coated to stop the fruit from browning.

Layer the fruit in the jars, and sprinkle the sugar over. Top with boiling water, ensuring all the fruit is submerged, and screw the lid on tightly. It's essential that the lids are well sealed to ensure no air can get in. Using a tea towel, gently turn the jar upside down 2–3 times to dissolve the sugar.

The jars of fruit will keep in the pantry for at least 6 months. Once you have opened a jar, store in the fridge and eat within 2–3 weeks.

03:   Cheese, milk and yoghurt

This chapter celebrates all that I love about cheese, milk and yoghurt, as well as some of my favourite plant-based alternatives including cultured cashew cheese, silky coconut yoghurt and toasted nut milk.

These simple, wholesome versions are actually a lot easier and less expensive to make than most people realise, and once you make your own, you might even feel inspired to experiment and add your own flair.

I've also included some lovely ideas on how to incorporate your homemade versions into more lavish dishes, such as herb-flecked yoghurt dip and macadamia milk custard.

Natural and coconut yoghurt

This luxuriously creamy and tangy yoghurt can be made with regular milk, or with coconut cream for a dairy-free option. Not only does this yoghurt taste thick, luscious and indulgent, it's also full of good, natural bacteria, which is beneficial for healthy gut flora.

Makes 1 litre

large thermos
1 litre full-fat milk or coconut cream
3 tablespoons plain natural yoghurt, or coconut yoghurt
2 tablespoons raw honey or maple syrup, optional*

Pour the milk or coconut cream into a large saucepan, and heat until just steaming, but not boiling, about 80°C. Allow to cool until it reaches 45°C.

While you are waiting for the milk to cool, preheat your thermos by filling it with boiling water. This creates a warm environment to start the yoghurt-making process. Once the milk or coconut cream has reached 45°C, stir in the yoghurt and honey or syrup, if using, and whisk gently to combine.

Pour out the water from the thermos, and then pour in the yoghurt mixture. Screw the lid on tightly, and then place in a warm area to culture for at least 12 hours, or up to 24 hours, depending on the weather and the consistency you are after.

Transfer the yoghurt to a clean glass jar, and chill in the fridge for a couple of hours. The yoghurt will thicken considerably during this time, and become thicker and slightly tarter each day, usually tasting best after 2–3 days. Store in the fridge for up to 2 weeks.

Be sure to save 3 tablespoons of yoghurt for the next batch.

The added sweetener is not absolutely necessary, although I find the bacteria loves to feed on the natural sugar, thus making a creamier yoghurt

Cultured cashew cheese

You could try this recipe using most types of nuts or seeds, however, I recommend using cashews to begin with, as they have a creamy, mellow flavour and blend easily. The culturing time depends on how warm your kitchen is and how sharp you prefer your cheese. Use your intuition, and have fun experimenting!

Makes 1 cheese

1½ cups raw cashew nuts,
    or other nuts or seeds of your choice
¼ cup extra virgin olive oil
¼ cup filtered water
1 teaspoon sea salt
1 teaspoon nutritional yeast
2 probiotic capsules*
1 small clove garlic
large piece of muslin or cheesecloth

To serve
Finely chopped herbs, nuts, seeds or spices for rolling

Soak the cashews for at least 6 hours or overnight, then drain and rinse well.

Place the cashews, along with the remaining ingredients, in a high-powered blender or food processor, and blend until completely smooth. This can take up to 5 minutes, and you might need to scrape down the sides a few times. Add a little more water, if needed, to help blend the mixture.

* These are the probiotic capsules you can buy from a chemist to help maintain a healthy gut

continues on next page

Place a sieve on top of a bowl, and line with a large piece of muslin or cheesecloth, allowing the edges to fall over the side. Carefully pour the cashew mixture into the sieve and gather together the corners of the muslin. Squeeze out as much excess liquid as possible, shaping the cheese into a flattened round about 3cm thick. Place the cheese in a round container or shallow bowl.

Put the cheese in a warm place to culture for 24 hours, and then place in the fridge for a further 10–14 days to firm up and ripen.

Once ready, the cheese should be firm to touch. Transfer to an airtight container and store in the fridge for up to 3 weeks.

To serve, gently roll the cheese in finely chopped herbs, nuts, seeds or spices.

Whole milk ricotta

If you haven't made your own cheese at home before, ricotta is a wonderful place to start as it doesn't require any fancy ingredients or equipment. I adore the gentle, milky texture dotted through a leafy green salad, or piled on top of toasted banana bread and drizzled with honey. The best-quality dairy will always give the best results, so I encourage you to choose organic, non-homogenised milk if you can. Once you've made up a batch, perhaps you'll enjoy my recipe for baked ricotta with thyme, orange and olives (page 67).

Makes about 1¾ cups

2 litres fresh whole milk*
1 teaspoon sea salt
80ml white vinegar
100ml filtered water
large piece of muslin or cheesecloth

Place the milk in a large saucepan over low heat, and stir until the milk starts to steam, around 90°C. Do not allow it to boil. Remove from the heat, and stir in the salt, vinegar and filtered water.

Allow the milk to rest for a few minutes, it should start to curdle and separate.

Place a sieve or colander on top of a bowl. Line the sieve with a large piece of muslin, allowing the edges to fall over the side. Using a slotted spoon, transfer the curds into the sieve, so that the whey drains while it cools.

You can use sheep's, cow's or goat's milk here

Transfer the ricotta to an airtight container. Store in the fridge, and enjoy within 3 days.

Baked ricotta with thyme, orange and olives

The soft and delicate flavour of ricotta is perfect for soaking up all kinds of delicious flavours, and on this occasion I've topped the cheese with thyme, orange zest and olives. This is a stunning addition to your cheese platter, especially alongside honey-scented oatcakes (page 102).

Serves 4–6 as a starter

1 cup whole milk ricotta (page 65)
¼ cup olives
2 tablespoons extra virgin olive oil
zest of 1 orange
small handful of fresh thyme
½ teaspoon dried chilli flakes
pinch of sea salt and pepper

Preheat the oven to 180°C.

Place the ricotta in a baking dish. Scatter over the olives and then drizzle with the olive oil and sprinkle with the orange zest, thyme, chilli flakes, and a pinch of sea salt and pepper.

Bake in the oven for 20–25 minutes until the ricotta is golden around the edges. Leave to cool and serve at room temperature.

Labne rolled in herbs and lemon zest

Here's a fabulous recipe for making your own tangy yoghurt cheese. These little creamy balls of delight are coated in fresh herbs and lemon zest, piled into a jar, and marinated in extra virgin olive oil. Quite the show-stopper, this is something to whip out when you have guests over.

Serves 6

2 cups natural or coconut yoghurt (page 59)
large handful of fresh herbs, finely chopped*
zest of 1 lemon
pinch of sea salt
pinch of ground pepper
about 1 cup extra virgin olive oil
large piece of muslin or cheesecloth

Over a large bowl, place a fine mesh sieve covered with a large piece of muslin or cheesecloth. Pour in the yoghurt, then twist together the corners of the cheesecloth like a moneybag.

Place a plate that fits inside the rim of the sieve, directly on top of the muslin-wrapped yoghurt, and then weigh it down with a can or something similar.

Refrigerate for at least 24 hours. The longer you leave the cheese, the more liquid will drip away and the drier the cheese will become.

Combine the herbs, lemon zest, sea salt and ground pepper in a shallow bowl. Using your hands, roll the cheese into small balls, and then roll in the herb mixture. Layer the balls in a large glass jar, then top with extra virgin olive oil. Secure the lid and refrigerate for up to a week.

Italian parsley, basil, mint, thyme and oregano all work well

Cashew sour cream

I just love the fresh, pure beauty of cashew sour cream with its soft tangy flavour and silky texture. Lovely served alongside crisp roasted wedges for dipping, or dolloped into a bowl of fragrant soup.

Makes about 1½ cups

2 cups raw cashews, soaked overnight
⅓ cup extra virgin olive oil
¼ cup apple cider vinegar
1 clove garlic

Place all the ingredients in a food processor or high-powered blender, and blend until completely smooth. This can take up to 5 minutes, and you might need to scrape down the sides a few times. Add a little more olive oil, if needed, to reach a sour cream consistency. Season to taste.

Transfer to an airtight container and store in the fridge for up to 5 days.

Herb-flecked yoghurt dip

Perfect for dolloping over all manner of things, this herb-flecked yoghurt will quickly become your new favourite dip. If you don't have fresh mint or chives, don't fret; you could also experiment with other herbs, such as parsley, basil, coriander or dill.

Makes 2 ½ cups

1 large telegraph cucumber
1¾ cups coconut or natural yoghurt (page 59)
handful of fresh mint, finely sliced
2 tablespoons extra virgin olive oil
2 tablespoons lemon juice
1 clove garlic, finely minced
½ teaspoon sea salt
pinch of cayenne pepper

Peel and finely slice the cucumber, then place in a bowl along with the remaining ingredients. Stir to combine, taste and adjust seasonings if necessary.

Wonderfully versatile, this fresh dip could be served alongside crispy grilled chicken or fish, as a bright creamy sauce in a pita bread, or perhaps dolloped on top of a hearty grain salad.

Store in a glass jar in the fridge for up to 4 days.

Toasted nut milk

The possibilities of this toasted nut milk are endless, really. Walnuts, almonds, hazelnuts, macadamia nuts – they're all equally wonderful, and give pleasantly different outcomes. The toasting of the nuts beforehand adds a deeper flavour to the milk, although it is not entirely necessary.

Makes about 1 litre

2 cups raw nuts
6 pitted medjool dates, optional
4 cups filtered water
¼ teaspoon sea salt

Toast the nuts in a dry skillet over medium heat, stirring every so often until fragrant and lightly golden, about 3–5 minutes. Keep a watchful eye towards the end as the nuts can burn easily.

Allow the nuts to cool, and then place in a large glass jar or bowl along with the dates, cover with filtered water, and allow to soak for at least 6 hours, or overnight.

Drain the nuts and reserve the (soaking) water. Place the drained nuts, dates and sea salt in a blender along with 1 cup of the water, and pulse until you have a creamy paste. Add the remaining water and blend until smooth.

Allow the milk to rest for 15 minutes, stir well, and then strain the milk through a fine mesh sieve. Transfer the milk to a glass bottle. You might like to reserve the leftover nut pulp to fold into a cake or muffin batter.

Store the milk in the fridge for 3 days. Simply tip upside down 1–2 times before using.

Vanilla bean mascarpone

Mascarpone is a gloriously indulgent Italian-style cheese that I like to think of as an elegant older sister to whipped cream. It's brilliant for stirring through savoury sauces to add a distinct rich flavour, as a filling for crêpes, or for creating a simple, summery dessert by layering with buttery almond crumble and glowing raspberry sauce. And despite seeming rather fancy, it's easy to make, too.

Makes about 2 cups

1 litre cream
1½ tablespoons white vinegar
1 teaspoon pure vanilla bean powder
large piece of muslin or cheesecloth

Make a double boiler by placing a large heatproof jug or stainless steel bowl over a simmering pan of water.

Pour the cream into the jug or bowl, and warm, stirring until it reaches 90°C. Stir in the vinegar and vanilla bean powder and continue to stir until the cream starts to curdle, about 2–3 minutes.

Remove from the heat and allow to cool for 10 minutes.

Place a sieve on top of a bowl, then line with a large piece of muslin, allowing the edges to fall over the side. Gently pour the cream into the sieve and place in the fridge for a day or so, allowing the cream to drain and thicken.

Discard the liquid from the bowl and transfer the mascarpone to an airtight container. Store in the fridge for up to 5 days.

Almond milk rice pudding with cardamom

Despite being rather old-fashioned, rice pudding is loved by most thanks to its lusciously creamy texture and comforting nature. Made using homemade almond milk and pure vanilla bean, this lovely pudding is elevated to perfection when topped with sweet preserved fruit (page 55) and crushed nuts.

Serves 6

200g Arborio rice
½ teaspoon ground cardamom
4 cups nut milk (page 75)
1 cup water
2 teaspoons pure vanilla bean paste

Place the rice, cardamom, milk, water and vanilla in a large saucepan over medium heat. Bring to a soft boil and then simmer on low for 30 minutes, stirring every so often until thick and creamy. You may need to add a little more milk towards the end.

To serve, divide the rice pudding between bowls and top generously with preserved or fresh fruit and crushed nuts.

Macadamia milk custard

This recipe calls for a touch of pure maple syrup – just enough to enhance the custard without making it overly sweet. The most joyous inclusion, however, is the addition of toasted macadamia milk, which gives this custard a luxuriously silky texture. Enjoy as it comes, or serve with sliced banana and berry sauce.

Serves 4

600ml macadamia milk (page 75)
¼ cup pure maple syrup
2 teaspoons pure vanilla extract
4 free-range egg yolks
1 tablespoon cornflour

To serve (optional)
sliced banana
berry sauce

In a bowl, whisk together the macadamia milk, maple syrup, vanilla extract, egg yolks and cornflour. Pour into a small saucepan over a low heat, and heat to a gentle simmer, whisking constantly so that the mixture slowly thickens. You'll know the custard is ready when it coats the back of a spoon.

Serve straight away, or transfer to an airtight container and refrigerate for up to 2 days.

Mint and chocolate frozen coconut yoghurt

Serves 4–6

The happiest of frozen delights, this bright and invigorating frozen yoghurt makes a wonderfully light dessert to eat. Fresh mint from the garden adds an unexpected herbaceous note, while chunks of dark chocolate make it luxuriously delicious.

1 cup natural or coconut yoghurt (page 59)
1 cup coconut cream
½ cup good-quality dark chocolate, finely sliced
¼ cup runny honey
1 tablespoon apple cider vinegar
large handful of finely chopped mint

Place all the ingredients in a bowl and mix gently to combine.

Tip the mixture into a container lined with baking paper, then place the lid on top. Freeze overnight, or for at least 8 hours.

Allow to soften for 5–10 minutes before serving.

04: Snacks, sweets and baked goods

If you're going to snack, you might as well do it well, and these quietly nutritious recipes are a delicious way to do so. In this chapter, you'll find a collection of wholesome bakes and snacks, including the best seeded crackers and chocolate buckwheat bumper bars, as well as the most marvellous berry yoghurt pancakes which I often pop into the kids' lunchboxes.

You'll also be pleased to know that these healthier, homemade versions won't have you spinning from a crazy sugar high as I've used lighter sweeteners, including honey, maple syrup and coconut sugar. Perfect for filling a gap, these recipes are sure to keep you happy, satisfied and full of energy.

Sweet potato hash browns

This is my healthy (yet infinitely delicious) take on the classic hash brown, using sweet potato for a lighter touch. The idea of salty, tender potato, grilled to crispy perfection, speaks to my soul like nothing else on a lazy weekend morning. Hash browns are also wonderful to pop into school lunches, which is why I've included them in this section.

Makes about
16 mini hash browns

3 cups grated sweet potato
3 free-range eggs
3 heaped tablespoons gluten-free flour (page 35)
   or spelt flour
1 teaspoon sea salt
coconut oil or golden ghee (page 23), for frying

Place the grated sweet potato between two paper towels and press well to soak up any excess liquid. This will ensure you have a crisper hash brown.

Whisk together the eggs, flour and salt until smooth, then fold in the grated sweet potato, and mix until well combined.

Grease a large pan, and place over high heat. Once hot, reduce the heat to low, then dollop tablespoons of the mixture into the pan. Cook for about 5 minutes on each side, until crisp and golden.

Repeat with the remaining mixture.

Heavenly served alongside eggs, or fancied up with toppings such as avocado, chilli flakes, a squeeze of lemon, and a drizzle of extra virgin olive oil.

Dukkah-crusted kale chips

These addictive kale chips don't last long in our household. In fact, an entire tray has been known to disappear within minutes of pulling it out of the oven. Salty, crisp and aromatic, these boisterously delicious chips are also tremendously good for you. They'll last for up to a week in an airtight container – if they don't get gobbled up, that is.

Serves 3–4 as a snack

large bunch of kale
extra virgin olive oil, for drizzling
dukkah (page 31), for sprinkling

Preheat the oven to 180°C, and line a large tray with baking paper.

Using a knife, carefully remove the leaves from the thick kale stems, and tear into smaller pieces.

Lay the kale on the prepared baking tray and drizzle lightly with extra virgin olive oil. Using your hands, gently massage the kale to ensure it is nicely coated in the oil, then sprinkle lightly with the dukkah.

Bake until the edges are crisp and slightly browned, about 10–15 minutes. Be sure to watch the kale carefully towards the end, as it can burn easily.

Remove from the oven and allow to cool on the tray.

Store in an airtight container for up to a week.

Super seeded crackers

Who doesn't love a tasty cracker? Particularly when they're perfectly salty, crunchy and delectably addictive. These are loaded with seeds, nuts and spices, and they're super popular with the little ones. I also love serving them alongside my tahini and cumin dip with honeyed eggplant (page 129).

Makes about
16 crackers

2 cups quick-cook rolled oats
1 cup sunflower seeds
½ cup pumpkin seeds
½ cup ground almonds
½ cup sesame seeds
½ cup chia seeds

2 tablespoons honey or
　coconut sugar
2 tablespoons olive oil or
　melted coconut oil
1 teaspoon sea salt
2½ cups water

Preheat the oven to 180°C, and line two baking trays with baking paper.

Combine all the ingredients in a large mixing bowl, and stir until well combined. The mixture will become thick and gluggy.

Divide the mixture between your two lined trays, and smooth out with the back of a spoon. Lay another piece of baking paper on top, and, using a rolling pin, roll out the mix to a thin paste, about ½cm thick. Remove the top piece of baking paper, and score the dough into rectangles.

Bake for about 25 minutes or until the crackers are lightly golden, crisp on the edges and snap apart easily. Be sure to check every few minutes towards the end as they burn easily. If they are still slightly soft, turn the oven off, and leave the crackers in there to firm up.

Store in an airtight container for up to 2 weeks.

Spelt and spinach wraps

Here's a wonderful recipe for spinach wraps. Soft and light to eat, I love to fill them with grilled chicken, salad and luscious mayonnaise (page 131) for a laid-back summer picnic. You can also make them ahead of time and pop them in the freezer.

Makes 8 wraps

2 cups fresh spinach leaves
1½ cups spelt flour
½ teaspoon finely ground sea salt
1 tablespoon olive oil, plus additional for cooking
½ cup hot water

Fill a medium-sized saucepan with water, and place over medium heat until simmering. Add the spinach leaves, and cook for a minute or so, until the leaves start to wilt. Drain the spinach using a colander, and rinse under cold water.

Finely slice the spinach, then place in a large mixing bowl along with the flour and salt. Stir in the oil and hot water a little at a time, and mix using a fork until the dough starts to come together.

Knead the dough on a floured surface until soft and elastic in texture, about 5 minutes.

Divide the dough into 8 pieces and shape each piece into a ball. Cover with a damp tea towel, and allow to rest for about 10 minutes.

Roll each dough ball into a circle, nice and thin like a tortilla should be, keeping in mind the size of your pan.

Grease a large pan and place over high heat. Reduce the heat to low, and then cook each tortilla until lightly golden, about 30 seconds on each side.

Repeat with the remaining dough.

Will keep in the fridge for up to 3 days.

No knead oat flour bread

Sometimes I just feel like making a simple bread that doesn't require kneading or fussy instructions, and I'm happy to say that this is the recipe for that. Sweet and mellow tasting, this oat flour bread has a soft, springy texture. I love how quickly it comes together, and the best part is that it only requires an hour of rising time. There's no need to feel wedded to the specific ingredients as it's a rather forgiving recipe – feel free to experiment with alternative flours such as rye, buckwheat or regular flour.

Makes 1 loaf

1½ cups warm water
2 teaspoons dried yeast
3 tablespoons honey or maple syrup
2¼ cups wholemeal spelt flour, or flour of your choice
1 cup rolled oats, plus 2 tablespoons for dusting
1½ teaspoons sea salt

Grease a 22 × 12cm loaf tin with coconut oil, melted butter or olive oil.

Combine the water, yeast and honey or maple syrup in a small bowl and stir together lightly. Leave to sit for about 10 minutes, so the yeast blooms and becomes foamy.

continues on next page

Place the remaining ingredients in a large bowl, reserving 2 tablespoons of oats. Add the yeast mixture and, using a fork, mix until the dough starts to come together. It will be quite sticky. Turn the dough into the prepared loaf tin, cover with a damp tea towel and leave to rise in a warm place for 1 hour.

Meanwhile, preheat the oven to 180°C.

Sprinkle the remaining 2 tablespoons of oats over the risen loaf. Bake the bread in the oven for 45 minutes, or until it is slightly browned on top and sounds hollow when tapped. Remove from the oven and turn out onto a cooling rack. Leave to cool for 20 minutes before slicing.

The bread will keep for up to 3 days in an airtight container and can be frozen for up to 2 months – slice before freezing for easy toasting.

Herb-flecked pizza dough

This astonishingly easy pizza dough has a lovely touch of fresh, earthy herbs. I roll the dough out nice and thin, brush it with garlic-infused extra virgin olive oil and then bake it at high heat until lightly charred on the edges. Best enjoyed warm, straight from the oven.

Makes 4 pizzas

2 cups warm water
2 teaspoons dried yeast
1 tablespoon coconut sugar
3 tablespoons extra virgin olive oil
5 cups spelt flour
½ cup finely chopped Italian parsley
1 teaspoon sea salt

Combine water, yeast and coconut sugar in a large bowl. Allow the yeast to bloom for 10 minutes, then, using a fork, stir in the olive oil, flour, parsley and sea salt.

Turn the dough out onto a floured bench top, and knead for 5–10 minutes until the dough is smooth and elastic.

Shape the dough into a ball, and place into a large bowl that's lightly coated with olive oil. Cover with a damp tea towel, and allow to rise for 2 hours in a warm spot.

Once the dough has doubled in size, punch it down and divide it into 4 equal-sized pieces. The dough is now ready to be used, or it can be popped in the freezer until you'd like to use it, for up to 3 months.

Preheat the oven to 220°C. Roll out the dough thinly, top with your favourite toppings and bake for 12–15 minutes, until crispy on the edges.

Sweet potato almond bread

I make a loaf of this almond bread every week as it makes me feel so much more energised and vital than regular bread does. Soft, light and sustaining all at once, it toasts beautifully, and makes a welcome foundation for an abundance of toppings. I sometimes swap out the sweet potato for pumpkin or potato, so you can imagine it's a wonderful recipe to use up any lingering roots in your pantry.

Makes 1 loaf

1 cup cooked sweet potato mash
1½ cups ground almonds
4 free-range eggs
¼ cup olive oil or melted coconut oil
¼ cup maple syrup
½ teaspoon sea salt
1 teaspoon baking soda
1 teaspoon apple cider vinegar
⅓ cup mixed seeds

Preheat the oven to 160°C, and line a 22 × 12cm loaf tin with baking paper.

Combine all the ingredients except the seeds in a large bowl, and mix until well combined. Pour the mixture into the prepared loaf tin, and smooth out evenly. Scatter the seeds over the top.

Bake in the oven for 1 hour, or until a skewer inserted into the middle of the loaf comes out clean.

Allow to cool in the tin for about 15 minutes, before turning out onto a cooling rack.

The bread will keep in an airtight container for up to 3 days and freezes well for up to 3 months.

Honey-scented oatcakes

These honey-scented oatcakes are perfect for topping with all manner of things, such as a thick slice of soft, creamy cheese and some chopped fresh figs. Crisp and buttery, you'll find these tender-crumbed biscuits are as delicious sweet as they are savoury.

Makes about
20 oatcakes

1 cup wholemeal spelt flour, plus more for dusting
1 cup rolled oats
¼ teaspoon baking soda
1 teaspoon sea salt
120g chilled unsalted butter, cubed
2 tablespoons honey
1 free-range egg

Preheat the oven to 180°C. Line two baking trays with baking paper.

To make the dough, combine the flour, rolled oats, baking soda, salt and butter in a food processor and pulse until the mixture resembles breadcrumbs.

Add the honey and egg, then continue to pulse until the dough comes together into a ball. Place the dough in the fridge to rest for 10–15 minutes.

Roll the dough out on a well-floured surface to about 5mm thick. Using a round cookie cutter (or the rim of a glass or jar), cut out as many circles as you can.

Transfer the oatcakes to the lined trays, leaving 2cm gaps between each. You might like to use a spatula to make this easier.

Bake for 8 minutes, or until lightly golden.

Remove from the oven and allow to cool on the baking trays. They can be stored in an airtight container for a week or so.

Cinnamon-baked pear chips

I just adore this recipe for baked pears, enhanced with a sprinkle of sweet, musky cinnamon, and baked until crisp. Try them with a sharp, salty cheese such as blue vein, or a gentler, creamy cheese like double cream Brie or Camembert.

These crunchy little chips are, of course, a satisfyingly delicious snack all by themselves. Great for filling lunch-boxes, too.

4 slightly under-ripe pears
3 tablespoons melted coconut oil
1 tablespoon pure maple syrup
1 teaspoon pure vanilla extract
1 teaspoon cinnamon

Preheat the oven to 100°C, and line two trays with baking paper.

Using a mandolin or sharp knife, slice the pears lengthwise into thin pieces. If doing by hand, try and get the slices as uniform in thickness as you possibly can.

Whisk together the coconut oil, maple syrup, vanilla and cinnamon to make a spiced syrup.

Paint each pear slice with the syrup, and then arrange the slices on the baking trays, in a single layer.

Bake for an hour, or until the edges begin to ruffle. Turn the slices over and continue to bake until crisp, about another half an hour. If the chips are still a little soft, leave them in a turned off oven for a further 30 minutes or so to finish crisping up. Be sure to keep a close eye on them, as they can burn easily.

Store in an airtight container for up to 1 week.

Courgette, apple and orange mini cakes

A recipe that is both indulgent and virtuous, these courgette, apple and orange-scented mini cakes are irresistibly moist and beautiful to eat. Enjoy them as they come, or top with vanilla bean mascarpone (page 77).

Makes 12

⅓ cup natural or coconut yoghurt (page 59)
⅓ cup melted coconut oil
¾ cup coconut sugar
2 free-range eggs, beaten
½ cup grated apple
½ cup grated courgette

zest of 1 orange
1½ cups spelt flour, or all-purpose gluten-free flour (page 35)
1 teaspoon baking soda
1 teaspoon apple cider vinegar
¼ teaspoon sea salt

Preheat the oven to 160°C, and lightly grease a 12-hole mini loaf tin or muffin tin.

In a large bowl, whisk together the yoghurt, coconut oil and coconut sugar. Fold in the remaining ingredients, and mix until well combined.

Divide the mixture amongst the prepared tins.

Bake in the oven for 20–25 minutes, or until a skewer comes out clean when inserted into the middle of each loaf.

Allow to cool for 5 minutes in the tins, and then carefully turn out onto a cooling rack.

Will keep in an airtight container for up to 3 days.

A simple nut flour cake for all seasons

Light, with a soft, sweet crumb, this nut flour cake is a brilliantly basic recipe, which can be topped with all kinds of seasonal fruit. It's simple in appearance, yet I still find it super gorgeous to look at. And, yes, billowy whipped cream would be a lovely accompaniment.

Serves 8

½ cup natural or coconut yoghurt (page 59)
¾ cup light muscovado sugar
½ cup olive oil or melted coconut oil
3 free-range eggs
2 teaspoons vanilla bean paste
1½ cups nut flour*

½ cup oat flour
1 teaspoon baking soda
1 teaspoon ground cinnamon
½ teaspoon sea salt

For the topping
seasonal fruit, sliced and cored
3 tablespoons oats or chopped nuts

Preheat the oven to 180°C, and line the base and sides of a 22cm cake tin with baking paper.

In a large bowl, whisk together the yoghurt, sugar, olive oil, eggs and vanilla, then gently fold in the dry ingredients until just combined.

Pour the batter into the cake tin and smooth out evenly. Arrange your fruit on top, and then sprinkle over the oats or chopped nuts.

Bake for 40 minutes, or until a skewer inserted into the centre of the cake comes out clean. Allow the cake to cool in the tin for about 15 minutes before carefully turning out onto a rack.

Almond, hazelnut or walnut flour/meal works well

Will keep in an airtight container for up to 4 days.

Triple coconut waffles

In this recipe, the classic waffle gets a wholesome spin, using simple and nourishing ingredients that you can feel better about biting into. I've switched out regular flour for spelt, and the natural sweetness comes from the coconut, honey and vanilla bean.

Makes about 5 waffles

1½ cups spelt flour*
1 cup coconut milk
½ cup desiccated coconut
3 tablespoons honey or pure maple syrup
3 tablespoons coconut oil, melted
3 free-range eggs
2 teaspoons pure vanilla extract
1 teaspoon baking soda

Place all the ingredients in a blender, and blend for 1–2 minutes until smooth and creamy. Place in the fridge for at least half an hour – this helps the batter to thicken. You can also pre-make the batter the night before so that it's ready for the morning.

Preheat the waffle iron and spray with coconut oil or neutral olive oil.

Pour ⅓ cup batter onto the centre of the waffle iron or pan, and cook until golden and lightly crisp. Repeat with the remaining batter.

You could also use all-purpose gluten-free flour mix (page 35)

Serve the waffles warm with vanilla berry syrup (page 163), and a generous dollop of coconut yoghurt (page 59).

Chocolate, buckwheat and fig bumper bars

These lusciously textured fig and buckwheat bumper bars are what I turn to when I'm looking for a delicious pick-me-up or sweet end to a meal. They're effortlessly easy to make – a heavenly combination of gooey dates, good-quality cocoa, coconut oil and ground almonds. I also love how the addition of buckwheat adds a pleasingly intriguing crunch. Cut into thick, decadent slices, and eat with joy.

Makes 8 bars

1 cup ground almonds or ground nuts of your choice
1 cup pitted fresh dates
¾ cup good-quality cocoa or cacao powder
½ cup dried figs, roughly chopped

2 teaspoons vanilla bean extract
pinch of sea salt
1 cup buckwheat groat
½ cup melted coconut oil
¼ cup cacao nibs

Line a 22 × 12cm loaf tin with baking paper.

Place all the ingredients in a food processor, except for the coconut oil and cacao nibs. Process for 20 seconds or so, or until the mixture resembles coarse crumbs.

Add the coconut oil and continue to pulse until the mixture starts to come together.

Press the mixture evenly into the lined tin, sprinkle the cacao nibs on top, and press down lightly.

Pop in the fridge for at least 2 hours to set.

Once ready, slice into thick bars and store in an airtight container in the fridge for up to 2 weeks.

Breakfast nut and seed truffles

School mornings can often be chaotic, which is why I created these satisfying little breakfast truffles to grab on the way out the door if I'm short on time. I'll often make up a batch on the weekend in preparation for the week ahead, when the need for something quick and healthy to eat will doubtlessly appear. Feel free to get creative with the seeds and nuts involved.

Makes about 16 truffles

1 cup ground almonds
½ cup sesame seeds
½ cup pumpkin seeds
4 tablespoons almond butter
4 tablespoons melted coconut oil
3 tablespoons water
1 tablespoon honey or pure maple syrup
1 teaspoon pure vanilla extract
pinch of sea salt
½ cup desiccated coconut, for coating

Place all the ingredients (except the desiccated coconut) into a food processor and process for 20 seconds or so, until the mixture starts to come together.

Using your hands, form the mixture into small balls, and then roll in the desiccated coconut until nicely coated.

Transfer the truffles to a plate, then place in the fridge for an hour to set.

The truffles can be kept at room temperature in an airtight container, but I find they are best kept in the fridge. They also freeze really well.

Raw brownies with salted nut butter caramel

Rich, gooey and impossibly decadent, these raspberry-topped raw brownies are my idea of perfection. Despite looking rather fancy, they're actually very easy to make. I've topped them with freeze-dried raspberries for a surprising hint of tartness, however coconut flakes or crushed nuts would also decorate them nicely.

Makes 12 small squares

For the brownie
1 cup fresh dates, pitted
1 cup buckwheat groats or rolled oats
¾ cup almonds or nuts of your choice
½ cup cocoa or cacao powder
pinch of sea salt
⅓ cup melted coconut oil
2 tablespoons water

For the caramel
¾ cup almond butter or
    nut butter of your choice (page 53)
½ cup coconut oil, melted and cooled
⅓ cup coconut cream
⅓ cup pure maple syrup or honey
2 teaspoons pure vanilla extract
pinch of sea salt

For the topping
1 cup freeze-dried or fresh raspberries

continues on next page

Line a 22 × 12cm loaf tin with baking paper.

Place the fresh dates and dry brownie ingredients in a food processor, and process for 20 seconds or so, until the mixture resembles coarse breadcrumbs. Add the coconut oil and water, and pulse until the mixture starts to come together. Press mixture evenly into the lined slice tin.

Place the caramel ingredients in a food processor or blender, and blend until silky and smooth. If the mixture starts to split, add some more coconut cream, a little at a time, scraping down the sides, until it comes together smoothly. This can sometimes happen quite quickly depending on the speed of your blender.

Pour the caramel on top of the brownie, and smooth out evenly using a spatula. Sprinkle the raspberries on top, and then pop in the freezer to set for at least 45 minutes.

Store in an airtight container in the freezer for up to 3 months. To serve, remove from the freezer and allow to thaw for a few minutes. Slice into small squares and enjoy.

Raspberry and yoghurt pancakes

Everybody needs a fabulous recipe for pancakes in their life, and I'm proud to say this one's mine. They're easy to whip up on a weekday morning, and those left over make a fantastic snack. If you don't have raspberries, don't fret, simply use other berries instead or press on without them.

Serves 4

2 cups wholemeal spelt flour*
1 cup natural or coconut yoghurt (page 59)
2 large eggs
4 tablespoons honey, maple syrup or brown rice syrup
1½ teaspoons baking soda
½ teaspoon sea salt
1¼ cups fresh or frozen raspberries
coconut oil or butter, for greasing the pan

To serve
yoghurt, honey, maple syrup or brown rice syrup on the side

Place all the ingredients in a blender, except for the raspberries. Blend until smooth. Gently fold in the raspberries.

Lightly grease a large pan over medium–low heat.

For each pancake, pour a heaping ⅓ cup of batter into the pan. As soon as little bubbles start to appear, turn the pancake over, then continue to cook for a further minute or so on the other side until golden. Continue with the remaining batter.

Serve the pancakes warm, with yoghurt, honey, maple syrup or brown rice syrup for drizzling.

* You could use buckwheat flour or all-purpose gluten-free flour (page 35)

Date, lemon and coconut yoghurt scones

There are so many ways to have fun with this tender, flaky scone recipe. You could switch out the dates and lemon for prunes, oranges and poppy seeds. Alternatively, vanilla and dried figs add a lovely texture to the dough. Chocolate chunks, apricots and a small pinch of spice would also be very, very good. You see, the possibilities are endless, really.

Makes 8

2 cups wholemeal spelt flour
2 tablespoons coconut sugar
½ teaspoon sea salt
1 teaspoon baking soda
zest of 1 lemon
60g hardened coconut oil or
   chilled unsalted butter

½ cup coconut or natural
   yoghurt (page 59)
1 cup pitted fresh dates,
   roughly chopped
nut milk (page 75),
   for brushing

Preheat the oven to 200°C, and line a tray with baking paper.

Combine the flour, coconut sugar, sea salt, baking soda and lemon zest in a large bowl, and whisk to combine.

Rub the coconut oil or butter into the mixture until it resembles coarse crumbs, then mix in the yoghurt and dates until you have a soft dough. If the mixture is a little dry, you can add a little more yoghurt until it comes together nicely.

Turn the dough out onto the baking tray, and shape into a large circle, about 3cm high. Mark into 8 triangles, scoring about halfway into the dough with a sharp knife, then brush the top with nut milk.

Bake in the oven for 10–12 minutes or until lightly golden and nicely risen. Allow to cool slightly, then enjoy warm.

05:   Sauces, dips and dressings

The following pages bring together my favourite sauces, dips and dressings that I use in my day-to-day cooking. Whenever I'm at a loss as to what to eat, I'll simply open up the fridge and grab a haphazard selection of bottles and jars, and in no time at all, I can muster up something wonderfully delicious to tuck into.

Once you've mastered a few of these simple preparations, you might feel inspired to combine them with whatever produce is in season, so that you can create new and exciting food for your table. There's no need to feel wedded to the specifics, have a play around with the recipes, and have fun doing so!

Tahini and cumin dip with honeyed eggplant

I love a dip with real character and depth, and this one certainly packs a punch. This is what I make if I have to take something to a friend's pot-luck dinner – it's a total crowd-pleaser and I know you'll also just love it.

1 eggplant, sliced in half lengthwise
3 tablespoons extra virgin olive oil
2 tablespoons runny honey
1 tablespoon cumin seeds
generous pinch of sea salt
1 heaped tablespoon tahini
juice of 1 lemon

To serve (optional)
olice oil for drizzling
basil leaves
1 tablespoon capers

Makes about 1½ cups

Preheat the oven to 180°C.

Place the eggplant in a roasting dish, drizzle with the extra virgin olive oil and honey, then sprinkle over the cumin seeds and sea salt.

Roast for 30 minutes, or until tender and caramelised. Remove from the oven and allow to cool for 5 minutes or so.

Place the eggplant in a food processor, along with the tahini and lemon juice. Process until smooth, taste, and adjust seasonings if you need to.

Serve with a drizzle of extra virgin olive oil, basil leaves and a sprinkling of capers if you desire.

Store in an airtight container in the fridge for up to 1 week.

Luscious mayonnaise

This creamy mayonnaise is rich in flavour and silky in texture. Delicious with chargrilled green vegetables, on a sandwich or as a salad dressing. You'll want to slather it on everything! Adding a touch of water to this recipe is a great little tip as it ensures you produce a thick, creamy and spreadable mayonnaise each time you make it.

Makes just under 2 cups

3 free-range egg yolks, at room temperature
2 cloves garlic, roughly chopped
1 tablespoon finely sliced preserved lemons (page 37)
½ teaspoon sea salt
1½ cups mild olive oil
1 tablespoon boiling water
1 tablespoon apple cider vinegar

Place the egg yolks, garlic, preserved lemons and sea salt in a blender.

Begin to blend the ingredients on low, pouring in the olive oil in a slow, steady trickle, followed by the boiling water and apple cider vinegar. You will know it's ready when the consistency begins to resemble a thick, glossy mayonnaise. This should take about 2–3 minutes.

Taste and adjust the seasonings if needed.

The mayonnaise will keep for 3–4 days stored in a jar in the fridge.

Roasted garlic and butter bean hummus

Preserved lemon and roasted garlic give a decent punch in this hummus, whilst the butter beans create an enticingly creamy texture. And, yes, it goes without saying that some toasted bread or crackers would be yummy for dipping.

Makes about 2 cups

1 bulb garlic, sliced in half
¼ cup extra virgin olive oil,
 plus extra for drizzling
400g can butter beans,
 drained
1 tablespoon finely sliced
 preserved lemon
juice of 1 lemon
flaky sea salt, to taste
ground pepper, to taste

To serve (optional)
olive oil for drizzling
1 teaspoon smoked
 paprika
cracked pepper

Preheat the oven to 180°C.

Place the garlic in a small roasting dish, flesh side up. Drizzle with extra virgin olive oil and roast until tender, about 20 minutes or so.

Allow to cool, then squeeze the flesh out of the papery skins, and place in a food processor along with the ¼ cup olive oil, butter beans, preserved lemon and lemon juice.

Process until you have a smooth, hummus consistency, and season to taste.

Serve drizzled with olive oil and sprinkle with smoked paprika and cracked pepper if desired.

Store in an airtight container in the fridge for 3–4 days.

Quick pickled red onions

Sweet, salty and sour, these pungent pickled onions are ridiculously easy to make and always come in handy. They're perfect for perking up a sandwich or piling on top of a roasted vegetable salad.

Makes 1½ cups

2 red onions
2 cups apple cider vinegar
2 heaped tablespoons coconut sugar
2 tablespoons extra virgin olive oil
2 teaspoons sea salt

Finely slice the red onions, then pack into a medium-sized jar.

Combine the apple cider vinegar, coconut sugar, extra virgin olive oil and sea salt in a saucepan over medium heat.

Bring to a gentle boil then carefully pour the mixture over the onions, ensuring the onions are completely submerged.

Allow to cool, then screw the lid on and refrigerate overnight. Your pickled onions will be ready to enjoy in about 8–12 hours and will keep in the fridge for up to 1 week.

Simple sauerkraut

Crunchy and delightfully sour, sauerkraut is inexpensive to make yet exceptionally wonderful to have on hand for adding another dimension to your plate. I love to pair it with a sharp Cheddar cheese inside a fancy toasted sandwich, or toss with lentils and plenty of fragrant herbs for a quick, punchy salad. Also fantastic served alongside slow-roasted meat.

Using only two ingredients, cabbage and sea salt, the fermentation process transforms this humble vegetable into one that is highly nutritious in vitamins and enzymes, as well as being rich in beneficial bacteria for the digestive system. You can add various types of herbs and spices to your kraut – a scattering of mustard, fennel or caraway seeds are my personal favourites.

Makes 2 × 400ml jars

1 large head cabbage
1 tablespoon sea salt
large glass jar

Remove the outer leaves from the cabbage, and then remove the core. Finely slice the cabbage, transfer to a large mixing bowl, and sprinkle the sea salt on top. Allow to rest for about 10 minutes so that the cabbage can soften and begin to release its juices.

Squeeze and massage the cabbage with your hands, releasing as much liquid as possible. This can be quite an arm workout and should take about 10 minutes or so.

Pack your cabbage, along with the juices, into a large glass jar, ensuring there are no air bubbles. Press the cabbage down until it is completely submerged in liquid, and then place something heavy on top to ensure it stays below the water line – a small jar filled with water works well. This will prevent mould from forming.

Place a piece of muslin or cheesecloth on top of the jar (leave the small, water-filled jar in as well), seal with a rubber band, and allow it to sit at room temperature for at least a week, and up to 4 weeks.

Check your sauerkraut daily, pressing it down if any cabbage might be floating above the brine.

Your sauerkraut will ferment faster in warm temperatures, and more slowly in winter. After about 2 weeks, begin to taste your sauerkraut periodically until it achieves the sourness you desire.

Once opened, store in the fridge and eat within a month.

Pistachio and mint pesto

This stunningly simple recipe combines pistachios, mint and garlic to create a gloriously fresh and zingy pesto. It is brilliant tossed with chunks of roasted cauliflower to create a punchy vegetable side, or used as a fabulous spread on toast with thick slices of creamy avocado.

Makes about 2 cups

1 cup raw pistachios
¾ cup extra virgin olive oil, plus extra if required
large handful fresh mint
large handful Italian flat-leaf parsley
2 tablespoons lemon juice
1 clove garlic

Dry toast the pistachios in a saucepan over medium heat until fragrant, then allow to cool for a few minutes.

Place the toasted pistachios, along with the remaining ingredients, in a food processor, and pulse until you have a chunky pesto consistency. Blend in more olive oil if you prefer a runnier consistency.

Season to taste.

Keeps for up to 1 week stored in a glass container in the fridge.

Sundried tomato chilli sauce

Sugar-packed chilli sauce gets an exciting makeover in this crazy-simple recipe which calls for fresh ingredients, rather than dried, for an ultimate flavour kick. Spoon it over boiled eggs, use it as a bright, spicy marinade for chicken, or use it to perk up a warming, vegetarian chilli dish.

Makes 1 × 300ml bottle

2 red chillies, deseeded and finely chopped
2 cloves garlic, finely chopped
3cm thumb of ginger, finely chopped
½ cup sundried tomatoes, finely sliced
½ cup apple cider vinegar
1 tablespoon honey or pure maple syrup
½ teaspoon sea salt
½ cup water

Combine all ingredients in a small saucepan over medium heat. Once boiling, reduce the heat to low, and simmer for 10 minutes until reduced by half, or until you have a syrup consistency.

Pour into a clean glass bottle and store in the fridge for up to 2 months.

Green dressing with capers and anchovies

This bright, creamy dressing is made from cashews and it's totally and utterly delicious. It'll magically transform even the most boring of salads, bring life to your sandwiches, or act as a luscious dip for roasted vegetables or crisp flatbread. In other words, you'll pretty much want to drizzle it over everything.

Makes 1½ cups

1 cup raw cashews, soaked in water overnight, then drained
¼ cup capers
2 fat anchovies
2 large handfuls of fresh parsley or basil
½ cup filtered water, plus more if needed
⅓ cup extra virgin olive oil
2 tablespoons apple cider vinegar or lemon juice
1 teaspoon pure maple syrup or 1 pitted date
1 heaped teaspoon Dijon mustard
1 teaspoon sea salt

Place all of the ingredients in a blender, and blend for 1–2 minutes, or until silky smooth. You can use a little more or less water, depending on your desired consistency. I start with about ½ cup and go from there. Check seasonings and adjust if necessary.

Pour into a glass jar and store in the fridge for up to 4 days.

My all-time favourite mustard vinaigrette

I love how a fantastic dressing can elevate the simplest of salads into something spectacular, and this mustard vinaigrette does just that. I'll always have a jar made up in the fridge. It is surprisingly delicious for dipping sourdough into – the bread soaks up all the mingling flavours in the best way possible so I encourage you to give it a go.

Makes about 1½ cups

⅓ cup extra virgin olive oil
¼ cup buttermilk or natural yoghurt (page 59)
juice of 1 lemon
1 tablespoon Dijon mustard
1 shallot, finely sliced

Combine all dressing ingredients in a small jar. Shake to combine, and then season with salt and freshly ground black pepper to taste.

Store in the fridge for up to 2 weeks.

Toasted walnut and parsley pesto

I use a variation of this herbaceous pesto in so many dishes that I simply can't remember when I first started using it. Use as a dip, toss through pasta, dollop into tacos or slather all over grilled fish. Most types of nuts and seeds will work wonderfully here.

Makes about 2 cups

1 cup raw walnuts
¾ cup extra virgin olive oil
6 anchovies, optional
2 large handfuls fresh parsley
2 tablespoons balsamic or apple cider vinegar
1 clove garlic

Dry toast the walnuts in a saucepan over medium heat until fragrant, then allow to cool for a few minutes.

Place the toasted walnuts along with the remaining ingredients in a food processor, and pulse until you have a chunky pesto consistency.

Season to taste, although not always necessary with the addition of salty anchovies.

Store in a glass container in the fridge for up to 1 week.

Cauliflower béchamel sauce

You'll be surprised to know this dairy-free take on classic béchamel sauce is made from cauliflower and cashew, yet it tastes very creamy and decadent. I can't imagine a sauce more luscious, melting and indulgently satisfying. This is a great recipe to breathe new life into leftover vegetables – simply roast the vegetables until tender, and then smother them in a blanket of delicious béchamel sauce. It's also fantastic layered in a classic moussaka.

Makes about 3 cups

½ head cauliflower, cut into florets and
   cooked until tender
1 cup cashews, soaked overnight
⅓ cup extra virgin olive oil
2 teaspoons Dijon mustard
2 cloves garlic
water, to blend

Place the cooked cauliflower, drained cashews, extra virgin olive oil, Dijon mustard and garlic in a food processor or blender. With the food processor running on low, gradually add water in a slow drizzle, until the sauce reaches a thick, silky consistency.

Store in the fridge in an airtight container for up to 3 days.

Smoky almond romesco sauce

I love the simple flavours here: roasted capsicum, smoky paprika, and a good hum of garlic to tie it all together. I like mine to be rustic and textured, however it's also sublime when blended until silky smooth. Toss it through pasta, dollop it on crackers or crispy fried eggs, or serve simply alongside a few hunks of delicious bread.

Makes about 2 cups

2 red capsicums or bell peppers, stems removed
1 cup toasted almonds
large handful of fresh parsley
⅓ cup extra virgin olive oil, plus extra for drizzling
1 tablespoon apple cider vinegar
1 teaspoon smoked paprika
1 clove garlic
1 teaspoon sea salt

Preheat the oven to 180°C.

Place the capsicum in a roasting dish, drizzle with extra virgin olive oil, and season with sea salt and cracked pepper.

Roast for 30 minutes, or until tender. Remove from the oven and allow to cool for 10 minutes.

Place the capsicums in a food processor, along with the remaining ingredients. Pulse until you achieve a rustic consistency, taste, and adjust seasonings if you need to.

Store in an airtight container in the fridge for up to 1 week.

Fragrant bone broth

You'll always find our slow cooker sitting on the kitchen bench as it holds steaming and fragrant bone broth for sipping on throughout the day. This nourishing broth contains plenty of vitamins and minerals, which are beneficial for your digestive health. You can also use leftover bones from a roast dish for a deeper, more aromatic flavour.

Makes about 10–12 cups

1kg organic chicken frames or beef bones
3 carrots, roughly chopped
3 celery stalks, roughly chopped
1 bulb garlic, each clove peeled and lightly smashed
1 leek or onion, roughly chopped
large handful of earthy herbs, such as thyme, rosemary or sage
2–3 bay leaves
1 tablespoon peppercorns

Place all ingredients in your slow cooker and cover with water.

Cook on a low setting overnight, or for at least 12 hours, then season to taste with sea salt.

Once the stock is ready, you can switch it to the low setting to keep it hot at all times for sipping on throughout the day. Alternatively, you can strain the stock through a sieve and transfer it to smaller containers.

The stock will keep in the fridge for up to 1 week, or in the freezer for up to 6 months.

Grandad's tomato sauce

I simply had to include this recipe for my grandad's legendary tomato sauce. I can guarantee it's unlike anything you've tried before, with its soft, subtle spice and slightly vinegary tang. This fabulous recipe is one to make when you have a serious glut of ripe tomatoes towards the end of summer. Note: you'll find the ingredients are measured in grams as this is how my grandad ensures the sauce turns out delicious every time!

Makes about
6 × 750ml bottles

3kg tomatoes, roughly chopped
25g garlic, peeled
1kg apples, roughly chopped
7 large brown onions, roughly chopped
110g sea salt
1kg sugar, I used light muscovado
500ml apple cider vinegar, plus 2–3 tablespoons
juice of 3 lemons
3½ tablespoons flour
2 tablespoons cornflour
large piece of muslin or cheesecloth

For the spice bag
28g whole peppercorns
15g whole allspice
15g whole cloves
15g cayenne pepper

continues on next page

Working in batches, place the tomatoes, garlic, apples and onions in a food processor, and pulse until finely chopped. Place the vegetables in a large stock pot over medium heat, along with the sea salt and sugar.

Enclose all spice bag ingredients in a handkerchief-sized piece of muslin or cheesecloth and tie tightly at the top, using a long piece of twine to lower the spice bag into the saucepan. Tie the string to the pot handle for easy retrieval. Bring to the boil, and simmer on low for at least 1 hour until the mixture is well softened.

Remove the spice bag and, working in batches, blend the sauce in a blender or food processor, until silky smooth. Pour the sauce back into the pot, and heat.

Add the vinegar and lemon juice (apart from adding to the flavour, the acetic acid in the juice is a natural preservative). Bring to the boil and simmer on low for about an hour. Keep stirring the sauce every 10 minutes or so.

In a small bowl, combine the flour and cornflour with enough vinegar to make a runny liquid. Remove the pot from the heat, and add the flour mixture to the sauce, stirring briskly. Put the pot back onto the heat and continue to stir until the sauce has thickened.

The sauce is now ready for bottling. Pour into sterilised bottles or jars, and screw on the lids tightly.

Will keep in the pantry for up to 5 years. Store in the fridge once opened.

Vanilla berry syrup

Sumptuously glossy, this simple berry syrup glows beautifully in a glass jar. The use of honey instead of sugar brings a mild, floral sweetness, while still tasting softly tart. Drizzle over ice cream, yoghurt or steaming oatmeal.

Makes about 2 cups

2 cups fresh or frozen berries
2 heaped tablespoons honey
1 teaspoon pure vanilla extract
½ cup water
1 tablespoon lemon juice

Combine the berries, honey, vanilla extract and water in a small saucepan over medium heat. Bring to a soft boil, and simmer for 15–20 minutes or until thick and glossy.

Stir in the lemon juice, then allow to cool for 10 minutes or so before transferring to a blender. Blend until lusciously silky and smooth.

Pour into a glass bottle or jar and keep in the fridge for up to 2 weeks.

Vegan chocolate fudge sauce

This is a killer chocolate fudge sauce for drizzling over all kinds of desserts. And because it contains coconut oil, it hardens when you pour it over ice cream, which will often create a flurry of excitement from kids and adults alike.

Makes 1½ cups

⅓ cup coconut oil, melted
¾ cup nut milk (page 75)
½ cup maple syrup
½ cup good-quality cocoa powder

Combine all ingredients in a blender and blend until silky smooth.

Pour into a glass bottle and store in the fridge for up to 1 month. It will thicken in the fridge, so you might like to bring it to room temperature or place it in a bath of hot water until it has the perfect drizzling consistency.

Rosemary-infused honey caramel sauce

Not only is this a beautiful sounding combo, it's also a deeply delicious one. So delicious, in fact, that I will confidently announce that it might be one of my favourite recipes in this entire book. I'll often whip up this caramel sauce to dress up a simple cake. I also need to warn you, it's dangerously addictive, so please keep this in mind when you make it.

Makes 1 cup

400ml coconut or cow's cream
½ cup honey or pure maple syrup
1 teaspoon pure vanilla extract
2 sprigs rosemary

In a saucepan over medium heat, combine all the ingredients, and bring to a soft boil.

Turn the heat down to low and simmer for 15–20 minutes, or until the caramel coats the back of a spoon. Stir often towards the end as it can burn easily.

Remove from the heat and carefully strain through a sieve into a clean jar.

Store in the fridge for up to 1 week. Bring the sauce to room temperature before serving.

Maple-sweetened lemon curd

This little jar of sunshine can bring so much glory to pancakes, waffles or your humble piece of toast. Slather it on with wild abandon and then thank yourself for making such a vibrantly delicious preserve. You could also make it using oranges or limes in place of lemons for something a little different.

Makes about 1 cup

⅓ cup melted coconut oil or unsalted butter
¼ cup pure maple syrup
2 free-range eggs, plus 2 egg yolks
zest of 2 lemons
½ cup freshly squeezed lemon juice

Put the coconut oil or butter and maple syrup in a small saucepan and melt together over medium heat. Remove from the heat and leave to cool.

Carefully whisk in the eggs, egg yolks, lemon zest and lemon juice.

Place back over a low heat, and continue to whisk for 2–3 minutes or until the curd becomes lovely and thick.

Remove from the heat, and allow to cool before pouring into a glass jar.

Store in the fridge for up to 1 week.

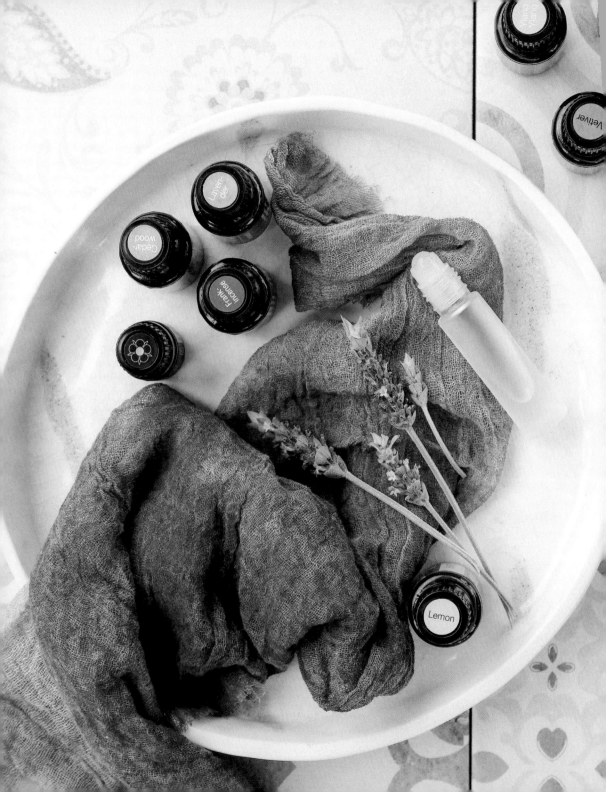

06: Beauty and wellbeing

When it comes to beauty and self-care, essentials oils have played a fundamental part in enhancing my wellbeing, in a way that feels effortless, natural and deeply empowering. I simply adore how they provide me the opportunity to live with more intention and care, not only for myself, but for those I love as well.

Welcoming essential oils into the rhythm of your day can be as simple as diffusing wild orange, lemon and bergamot to purify the air in your home, or massaging a diluted blend of frankincense and lavender into your shoulders to relieve stress. Perhaps you might enjoy whipping up your own divine-smelling beauty products that are satisfyingly simple and fun to make.

There were many recipes that I wanted to include in this chapter, however, I've managed to narrow it down to those I love most, including my daughter Bella's personal favourite – lavender bath melts. I hope you'll have fun creating them and breathing in their beautiful aromas, but above all, I hope they'll allow you to take more time to honour, love and take care of yourself.

The magic of essential oils

So what are essential oils? In a nutshell, they are the lifeblood of plants. Leaves, petals, stems, bark, resin, fruit and herbs are all beautiful parts of our natural world that produce incredibly powerful and highly concentrated oils; most of them are 50–70 times more powerful as oils than they are in natural plant form. Most are primarily extracted through careful steam distillation and cold-pressing, to ensure the quality and purity of the oil is untouched.

Unfortunately, not all bottled essential oils are created equal, and many contain fillers and artificial fragrances. It's for this reason that I encourage you to find 100 per cent pure, therapeutic grade oils to ensure you are receiving the best quality for you, your home and your family.

Diffuser recipes

One of the easiest and most popular ways to enjoy essential oils is by using an ionic diffuser. The aromatic scent of essential oils can support feelings of invigoration, clarity and peace, among other empowering emotions, as well as helping to cleanse the air in your home. Here's a selection of harmonious blends you might like to try for various seasons and moods.

Cleanse – grapefruit, lime and bergamot

Uplift – lime, lemon, grapefruit and wild orange

Purify – wild orange, Douglas fir and lemon

Refresh – tangerine, lemon and peppermint

Focus – lime, cedar wood, frankincense and bergamot

Energise – grapefruit, peppermint and lemon

Immunity – tea tree, frankincense and oregano

Breathe – eucalyptus, peppermint and lemon

Sleep – lavender, vetiver and cedar wood

Winter spice – wild orange, cinnamon, frankincense and clove

De-stress – lavender, frankincense and ylang ylang

Allergy relief – lavender, lemon and peppermint

Memory booster – rosemary, lemon and frankincense

Bug deterrent – tea tree and lavender

Perfume balm

Compose your very own signature scent with this blended perfume balm. It's wonderfully easy to prepare and can be made with just a handful of natural ingredients.

Your perfume balm can be as simple or as complex as you choose – it's all about developing your own personal scent profile that's unique to your senses. You can go with a single note fragrance, such as rose, bergamot or lavender, or you might like to try your hand at combining oils to come up with something unique.

Here's a few lovely combinations you might like to try:

* bergamot cedar wood and grapefruit

* lavender, wild orange and vetiver

* rose, sandalwood and jasmine

* bergamot, ylang ylang and vanilla absolute

* patchouli and wild orange

* neroli, jasmine and rose

* ylang ylang, tangerine and frankincense

* lemongrass, lime and vetiver

Note: some oils are not recommended to use during pregnancy. Please seek medical advice before using. Also be careful to avoid using any 'hot' oils on the skin, such as clove, cinnamon and oregano.

What you'll need
6 tablespoons grated beeswax
6 tablespoons fractionated coconut oil,
   almond oil or jojoba oil
essential oils of your choice
small glass cosmetic jars

Combine the beeswax and carrier oil over a double boiler,
and stir until melted.

Add 15–20 drops of the essential oils, smelling as you go along.

Once you are happy with the fragrance, carefully pour
the mixture into jars and allow to solidify, this usually takes
2–3 hours.

To use, rub the balm onto your wrists and neck as desired.

Make-your-own roller blends

Incorporate the magic of essential oils into your day with these easy-to-make roller blends. I always pop a few in my handbag and in the kids' school bags, so that we can whip them out whenever needed. The little ones also love to join in when I'm making them, and they like to create their very own combinations.

Because these oils are diluted with a carrier oil, they're safe to use on children from 12 months plus, and can be applied as often as needed. *Please be careful to avoid using any 'hot' oils on the skin, such as clove, cinnamon and oregano.*

Note: some oils cannot be used during pregnancy, or if you have any health conditions. Please seek medical advice before using.

What you'll need
10ml glass roller bottle
essential oils of your choice
   (see my favourite blends, page 173)
carrier oil*

Once you've decided on the blend you'd like to make, carefully place 8–15 drops of essential oils in your roller bottle. If you are making a blend for a child over 12 months, start with 2 drops and go from there.

* I use fractionated coconut oil as I find it absorbs into the skin easily, although almond oil or jojoba oil also work well

Fill to the top with the carrier oil of your choice, pop the lid on, and carefully tip upside down a few times to disperse.

To use, apply to the wrists, soles of the feet, along the spine, temples, back of the neck or your particularly affected area.

Ylang ylang and tea tree face serum

The exotic, floral scent of ylang ylang is the star of this luxurious face serum, perfectly complemented by lavender and frankincense, with tea tree to help reduce blemishes. I love using it as part of my simple beauty routine, as it gives my skin a soft glow and makes me feel like a goddess. Beautiful for both the face and body.

30ml fractionated coconut oil, almond oil or jojoba oil
8 drops ylang ylang essential oil
6 drops lavender essential oil
4 drops frankincense essential oil
4 drops tea tree essential oil

Combine all ingredients in a small dropper bottle, and gently turn upside down a few times to disperse.

To use, gently massage a few drops into your skin after cleansing.

Brightening pink clay mask

French clay is one of those magical beauty ingredients that truly is as wonderful as everyone claims. I was pleasantly surprised by the glow on my skin after first using it, and now I'm hopelessly hooked.

The wonderful thing about clay is that it naturally draws out toxins and impurities, and has a gentle yet effective nature. I've combined it with rose water and raw honey in this recipe, as both assist in healing and nurturing the skin. I find it is best made fresh with each use.

2 tablespoons rose water
3 teaspoons witch hazel
1 teaspoon raw honey
2 tablespoons green or pink French clay*
2 drops lavender essential oil

Stir together all the ingredients in a small bowl until you have a smooth paste.

Dampen a face cloth with warm water, and hold over your face for 15 seconds or so to open your pores.

Apply the face mask over your face and neck, avoiding eyes and hairline. Leave on for 15–20 minutes, and then rinse off with warm water. I recommend applying a few drops of ylang ylang and tea tree face serum (page 179) afterwards.

You can find this at most natural skincare and aromatherapy stores

Please note that your skin may be slightly pink for a little while afterwards, which is normal and caused by increased circulation.

Geranium face steam

Steam is the easiest way to help open your pores and soften your skin, which means your face oil and moisturiser can better permeate your skin. You don't need to visit a fancy day spa to indulge in a facial steam – you can treat yourself to one at home! It's so simple, but it feels oh-so-beautiful.

What you'll need
kettle filled with boiling water
heatproof bowl
towel

Essential oil combinations
Detox – 4 drops tea tree and 6 drops lemongrass
Balance – 5 drops geranium and 5 drops sweet orange
Relax – 4 drops lavender and 6 drops lemon

To start, cleanse your skin using a warm cloth to remove any makeup or impurities.

Bring the kettle to a boil, and carefully pour the steaming water into a bowl. Add your combination of essential oils.

Sit with your face over the bowl – being careful not to get too close to the scalding water – with a towel draped over your head. Steam your face for about 5 minutes, and no longer than 10 minutes. Breathe deeply, inhale, and feel beautifully relaxed.

Finish by using your favourite face oil or moisturiser, to help seal in all the goodness.

Natural deodorant spray with patchouli, tea tree and bergamot

This spray blend of patchouli, tea tree and bergamot not only smells heavenly, it works naturally to combat body odour and has antibacterial properties.

4 drops patchouli essential oil
6 drops tea tree essential oil
5 drops bergamot essential oil
40ml rose water

Makes 50ml

Put all the ingredients into a dark-coloured spray bottle. Shake gently to combine.

Store in a cool, dark place, and use as needed.

Gardeners' peppermint hand scrub

Plenty of time spent in the garden can leave your hands feeling rough and battered. This invigorating hand scrub will add much needed softness; and it smells incredible. It's also fantastic for your feet.

1 cup fine sea salt
¼ cup fractionated coconut oil or almond oil
5 drops peppermint essential oil
5 drops lime essential oil

Combine all the ingredients and store in a glass jar.

To use, gently exfoliate skin with 2 tablespoons of the scrub for at least 2 minutes, then rinse with warm water.

Will keep in a cool, dry place for up to 6 months.

Sea salt spray for beach hair

Lazy days at the beach combined with sea salt and sun can create a loose, carefree hairstyle that looks effortlessly easy. The good news is you can recreate that summer hairdo with this simple-to-make salt spray. I've included coconut oil to replenish your locks, and a lovely blend of lavender and lime essential oils so your hair smells incredible.

250ml glass spray bottle
1 cup warm water
2 tablespoons sea salt
1 tablespoon natural conditioner
1 teaspoon fractionated coconut oil, almond oil or jojoba oil
5 drops lavender essential oil
5 drops lime essential oil

Pour all ingredients into the spray bottle and shake gently to combine.

To use, apply to towel-dried hair, and scrunch using your hands to create texture. Allow to dry naturally.

Will keep in a cool, dry place for up to 3 months.

Healing hand salve with olive oil

One batch of this recipe makes three small amber jars of balm, so I can stash one in my purse as well as having a couple of extras in the bathroom cabinet. The nourishing shea butter helps to trap in the moisture and protect dry palms, while the olive oil adds a lovely, soft texture.

¼ cup olive oil
¼ cup shea butter
5 drops lavender essential oil
5 drops ylang ylang essential oil
5 drops frankincense essential oil
3 × 50ml amber jars

Place the olive oil and shea butter in a heatproof bowl over a double boiler. Stir until melted, then stir in the essential oils.

Pour into the amber jars and allow to set for about 2 hours.

The salve will keep in a cool, dry place for up to 6 months.

Coconut milk and peppermint bath soak

I'll use any excuse to take a long, hot soak in the bath. And this invigorating peppermint soak is particularly fantastic for boosting my mood. The coconut milk powder in this recipe adds a certain luxuriousness that wraps you up like a soft, warm blanket. You could switch out the peppermint for something more soothing, such as frankincense or lavender, if you'd prefer a more relaxing experience.

1 cup pink Himalayan or Epsom salts
1 cup coconut milk powder*
10 drops peppermint essential oil or oil of your choice

Put all the ingredients into a bowl and toss gently to combine. Transfer to a dark-coloured glass jar to store.

Keep the jar in a cool, dark place, and use within 1–2 months.

You could also use cow's milk or almond milk powder

To use, add a large handful to your bath and allow it to dissolve in the hot water.

Lavender and vetiver bath melts

Place a handful of these beautifully scented melts into your bathing waters and you'll be magically transported to a state of blissful repose. Lavender and vetiver essential oils are both renowned for their ability to soothe the senses and calm the soul, and when combined with shea butter and coconut oil, your skin will be left feeling nourished and oh-so-soft.

Makes 16

½ cup shea butter
1 tablespoon coconut oil
3 tablespoons grated beeswax
10 drops lavender essential oil
10 drops vetiver essential oil

Melt together the shea butter, coconut oil and beeswax over a double boiler, then add the essential oils, and stir until combined.

Pour into a silicon ice tray, and allow to harden for at least 4 hours. Pop out each melt, and then store in a glass jar or airtight container.

To use, place two or three into your bath, and watch as they melt beautifully into your water.

Epsom salt, lavender and cedar wood bath soak

Epsom salts are a brilliant source of magnesium, a mineral that can help with muscle aches, PMS and insomnia. Dissolving Epsom salts into your bathing waters is a simple way to boost your magnesium levels – it absorbs into your skin while you unwind and relax. If I'm feeling extra fancy, I'll add a handful of fresh petals to the bath water for a rejuvenating floral soak.

1 cup Epsom salts
6 drops lavender essential oil
4 drops ylang ylang essential oil
2 drops cedar wood essential oil

Place all ingredients into a bowl and toss gently to combine. Transfer to a dark-coloured glass jar to store.

Keep the jar in a cool, dark place, and use within 6 months.

To use, add a large handful to your bath and allow it to dissolve in the hot water. It's best to do this right before you hop in the bath as this is when the essential oils are most fragrant. Inhale deeply and enjoy.

07:   For the home

I first started making my own natural products for our home about 5 years ago, when I became obsessed with creating a healthier, less toxic living environment for our family. I've fallen in love with whipping up new sprays, cleaners and other various earth-friendly goods to keep our house smelling and looking beautiful, and it's a journey that continues to surprise and delight.

This chapter offers some of our favourite DIY methods, including the best all-purpose spray and beeswax food wraps. All are fantastic as a rainy day activity, and great for getting the little ones involved.

Naturally scented soy candles

Scented with a single note of vanilla bean, these natural soy candles are perfect in their simplicity. The pleasant glow of a flickering candle in the evening is such a simple pleasure, and even more gratifying when the candle has been made using your own hands.

What you'll need
grated soy wax*
glass jar or vessel
candle wick that is slightly longer than the height of your jar
15–20 drops of vanilla absolute essential oil, or essential oils
   of your choice
large peg

Pour boiling water into your glass jar to warm it up, then pour out the water. Once the jar is cool enough to handle, dry it thoroughly.

Secure the metal bottom of your wick onto the centre bottom of the jar using double-sided tape or a hot glue gun.

Measure the approximate amount of wax you'll need (depending on the size of your jar) into a glass measuring jug – you'll need double the amount as it melts down considerably.

* Use about 225g of soy wax for a 300ml jar

Melt the wax over a double boiler. Remove from the heat and stir in your essential oils.

Carefully pour the melted wax into the jar, using a peg to keep the wick in place while the candle sets.

Place it somewhere warm overnight to set, then trim the wick if needed.

The first time you light your candle, keep it alight for at least an hour, or long enough to allow the entire top of the candle to melt. This is to ensure even burning for future use.

Wild orange and clove air freshener

This zesty spray is brilliant for purifying the air, and is particularly good at diminishing any lingering odours in the kitchen. During the warmer months, I'll switch out the spicy clove for something a little more refreshing such as eucalyptus or grapefruit. Patchouli or geranium paired with wild orange is another lovely combination.

Makes 50ml

10 drops wild orange essential oil
5 drops clove essential oil
10ml witch hazel

Put all the ingredients into a dark-coloured spray bottle and top with filtered water. Shake gently to combine.

Store in a cool, dark place, and use as needed.

Purifying smudge sticks

Here's a beautiful, all-natural antidote to cleansing stagnant or negative energy within your home. You can choose to combine a variety of herbs and flowers, with my personal favourite being lavender, eucalyptus and rosemary. Once you've wrapped up your sticks, they'll need to dry out for a month or so. As a bonus, they do look rather pretty sitting on your windowsill.

Choose a selection of:

* thyme

* eucalyptus

* mint

* sage

* pine

* roses

* rosemary

* lavender

natural cotton or hemp string

continues on next page

Put together a bundle of your desired herbs and flora, much like you would arrange a bouquet of flowers.

Next, wrap the natural cotton or hemp string tightly around the base of your stems, tying a knot to secure the string. Continue to wind the string around the stems, wrapping them tighter than you might expect as they will shrink in size as they dry. Wind up and down the length of the stems 4–5 times until tightly bound together. Secure with another knot at the end.

Place the bundle in a warm, dry place such as a hot water cupboard, and allow to dry completely, for 3–4 weeks.

To use your smudge stick, light the flowering end and, as it starts to crackle, blow out the flame. Gently 'smudge' the aromatic smoke around the room, particularly in the stagnant corners. You might need to gently blow on the end a few times to keep it smouldering.

A few things to note:

Only light your smudge stick in a well-ventilated room, making sure it's not in close contact with any flammable items such as blankets or curtains.

Once you are finished with your smudge stick, douse it in water to stop it from smouldering, and then place in a heat-safe container made of ceramic or glass.

Do not leave unattended, or in reach of small children.

Essential oil reed diffuser

To make your own all-natural reed diffuser, all you'll need is a handful of basic ingredients, some bamboo reeds and a glass vessel with a narrow neck. I've chosen to share a harmonious blend of tangerine, cedar wood and frankincense, however I encourage you to experiment with different essential oil combinations as you desire. Hop on back to page 173 if you need some more ideas for lovely blends.

What you'll need
glass vessel with a narrow opening
¼ cup carrier oil, such as fractionated coconut oil,
    sweet almond or light olive oil
¼ cup witch hazel
10 drops tangerine essential oil
10 drops cedar wood essential oil
10 drops frankincense essential oil
8 bamboo reeds or skewers

In a measuring jug, place the carrier oil and witch hazel – this helps to thin out the oils, allowing them to travel up the reeds. Stir in the essential oils to combine.

Trim off the pointy ends on the bamboo skewers, bundle the reeds together and insert them into the neck of the vessel. Allow the sticks to soak up the oil for an hour or so, then flip them over.

I flip the bamboo sticks every few days, to keep the fragrance wafting.

Wool dryer balls

We all know that drying our laundry in the sunshine is the loveliest way to get the job done, but during the colder months it's inevitable that you'll need to make use of a tumble dryer. This is where these wonderful little balls come in extra handy.

Not only do wool dryer balls help to keep your laundry soft and knock out wrinkles, they're completely natural, being made from felted wool, and can last forever. Simply pop them in your dryer along with your clean washing. Your wallet (and the planet) will thank you.

Makes 6 balls

What you'll need
2 balls of 100 per cent wool yarn
old pair of stockings
large sewing needle

Begin by winding the wool around two fingers, about 15–20 times. Slip the wool from your fingers, and continue to wind the yarn around, moving the bundle in a slow circular motion so you start to form a ball shape.

Continue to add layers, until it's about twice the size you want – it will shrink down considerably. Once you are happy with the size, snip the yarn, and then, using a needle, sew the wool through the middle of the ball to the other side to ensure it doesn't unravel.

Repeat with the remaining wool until you have 6 balls.

Cut off a leg from your stockings, and place the balls in one at a time, securing each with a knot in between.

Place into a large saucepan of water, bring to the boil, and simmer on low for about 15 minutes.

Allow to cool slightly, and then pop them into your next load of washing to help the balls shrink and felt.

Once the load has finished, remove the stocking, and then pop the balls into the tumble dryer along with your clean, wet washing.

If you'd like to add some lovely scent to your next load, add 2–3 drops of essential oils to each ball to infuse your washing with a beautiful aroma.

The best all-purpose spray

This bright and zippy all-purpose spray will leave surfaces sparkling, and your home will smell delightful, too. Not only will this spray bring more joy to the task of cleaning, the citrus essential oils help to uplift your mood and awaken your senses.

Makes about 500ml

2 teaspoons baking soda
2 teaspoons liquid castile soap
10 drops bergamot essential oil
10 drops lime essential oil
10 drops lemon essential oil
475ml hot water

Place all the ingredients into a dark-coloured spray bottle and shake gently to combine before each use.

Lemon and lime dishwashing powder

Lemon essential oil has this amazing ability to cut through grease, making it the star ingredient in this extra-zesty dishwashing powder. The method is simple, and all you'll need is a handful of basic ingredients and a glass jar to mix it all up in.

Washes about
35–40 loads

1½ cups baking soda or washing soda
½ cup citric acid
⅓ cup fine salt
10 drops lemon essential oil
10 drops lime essential oil

Combine all the ingredients in a large glass jar, screw on the lid, and tip upside down 4–5 times to mix well.

Use 1 tablespoon per load as needed.

For an extra clean load, add white vinegar to the rinse compartment, and wash as usual.

Beeswax
food wraps

These homemade beeswax wraps are the prettiest way
to keep food fresh without the use of plastic.

Wrap fresh produce, cheese and sandwiches in them, and
they are also rather nifty for covering a jar or bowl. The
best part? The wraps are reusable! Simply rinse under warm
water and allow to air dry. I've been using mine for about a
year now, and they're still going strong.

What you'll need
beeswax, either in block form or in beads
cotton fabric – I find patterned is best as the wax leaves a
    yellow tinge which doesn't look so pretty on plain colours
an old piece of fabric, which you don't mind getting wax on,
    to use as a tablecloth during the process
sharp scissors (or pinking shears if you're fancy like that)
wax paper
an iron

Start by cutting your fabric into the shapes/sizes that
you desire. Ensure the sizes will fit within the confines of
your wax paper.

Lay the tablecloth down, and switch on your iron to warm it
up. I put mine on the cotton setting.

Lay down a piece of wax paper, and place a piece of fabric
on top. If you are using beeswax in block form, you will need
to grate it. Sprinkle the grated beeswax or beads evenly on
top of the fabric – you don't need a huge amount, start with
less and see how you go.

Place another piece of wax paper on top of the wax and fabric, and iron gently until the wax melts evenly. Carefully peel off the wax paper while the wax is still warm, then wave the fabric around in the air for a few seconds while the wax hardens. Drape or hang the new beeswax food wrap over something to dry (I use a wooden dish rack). Repeat until you have used up all of your fabric.

If the edges need neatening up, you can trim them once the wrap has cooled down.

Recipe Index

Household Index

Acknowledgements

I absolutely loved working on this book, and not only because of its subject, and how easily the recipes flowed, but also because of all the beautiful people who so generously contributed in some way or another. I'm so appreciative, and indebted to you all.

Firstly, a huge thank you to the wonderful team at Penguin Random House; in particular, Debra, Jeremy, Sarah and Rachel. Thank you for sharing my vision of *Homemade* and for bringing my dream to fruition. Isn't it utterly stunning?

There are plenty of individuals who, in many ways, helped and inspired me to create this book. Morrae from Boutique Barn, thank you for opening your home and lovely venue space for us, and despite the rain, helping us create such a romantically rustic setting. Yelena Bebich, my fabulous makeup gal, you sure do know how to make me feel extra-sparkly!

Much love to Julz, who helped to share the photography load. I've so enjoyed working with you over the past few years. How amazing that both our style and aesthetic seem to just click so beautifully. I couldn't have done this without you!

Big thanks to Hej-Hej and Ellis labels for gifting me some lovely outfits to wear, and Thea Ceramics, Dawn Clayden, and Bohome and Roam for the exquisite pottery and linens which you'll find dotted throughout these pages.

To my loyal readers, thank you for holding space for me, and for years of support and encouragement. What a privilege it is to share my love of food and simple living with you all!

And last but not least, Valentin, Bella, Obi and Archie, my angels! I love you so much, and do this all for you.

Eleanor x

About the author

Eleanor Ozich is a bestselling author and photographer and mother to three young children. Since starting a blog six years ago, she has built a large following of loyal readers who look forward to her daily musings, recipes and ideas for living a less complicated way of life. Her unique approach to writing and photography showcases her love of all things simple in a natural, down-to-earth way.

A self-taught cook, Eleanor grew up living above her parents' restaurant in Auckland. She has published two cookbooks, *My Petite Kitchen* and *My Family Table*, a lifestyle guide, *The Art of Simple*, and contributes to various publications, including *Taste* magazine and *The Natural Parent Magazine*. She is also a contributor to *Viva* in the *New Zealand Herald*.

Eleanor and her family live in west Auckland, surrounded by rolling green hills.

Also by Eleanor Ozich

PENGUIN

UK | USA | Canada | Ireland | Australia
India | New Zealand | South Africa | China

Penguin is an imprint of the Penguin Random House group of companies, whose addresses can be found at global.penguinrandomhouse.com.

 Penguin
Random House
New Zealand

First published by
Penguin Random House New Zealand, 2019

1 3 5 7 9 10 8 6 4 2

Text and photography © Eleanor Ozich, 2019

The moral right of the author has been asserted.

Design by Rachel Clark ©
Penguin Random House New Zealand
Photographs on pages 7, 28, 32, 37, 38, 42, 48, 56, 63, 105, 108, 130, 152, 153, 166 (and cover), 186, 187, 206 and 209 by Julia Glover
Prepress by Image Centre Group
Printed and bound in China by RR Donnelley

A catalogue record for this book is available from the National Library of New Zealand.

ISBN 978-0-14-377280-4

penguin.co.nz

FSC
www.fsc.org

MIX
Paper from
responsible sources
FSC® C101537